Rasul Ahtyamov
Natalia Meshcheryakova

Reduction of greenhouse gas emissions boiler plants

Rasul Ahtyamov
Natalia Meshcheryakova

Reduction of greenhouse gas emissions boiler plants

ScienciaScripts

Cover image: www.ingimage.com

This book is a translation from the original published under ISBN 978-620-5-51994-3.

Publisher:
Sciencia Scripts
is a trademark of
Dodo Books Indian Ocean Ltd. and OmniScriptum S.R.L publishing group

120 High Road, East Finchley, London, N2 9ED, United Kingdom
Str. Armeneasca 28/1, office 1, Chisinau MD-2012, Republic of Moldova, Europe
Printed at: see last page
ISBN: 978-620-6-23307-7

Contents

Introduction 2
CHAPTER 1 3
CHAPTER 2 11
Bibliography 66

Introduction

An important problem is the high concentration of pollutants, which mainly affects the health of workers, as well as the health of the neighbouring population. In addition, pollutants have a negative impact on the climate.

Climate change due to greenhouse gas emissions is becoming a global problem. In 2016, the concentration of carbon dioxide in the Earth's atmosphere reached a record high of 400 ppm (parts per million - the number of CO2 particles per million air particles). And by the end of the century, the concentration of CO2 is expected to increase approximately twofold. Despite the targeted growth of solar and wind energy, there is still no competitive alternative to burning hydrocarbon fuels. In this context, carbon capture and storage technology (CCS) is recognised as critical for limiting global temperature rise to between 1.5 and 2^{0} C by 2050. The application of these technologies will reduce the amount of co2 emitted by businesses. Thus, this problem is urgent from the resource-saving, environmental and economic points of view and requires implementation at enterprises

CHAPTER 1

ANALYSING THE CURRENT STATE OF THE SAFETY PROBLEM ON THE EXAMPLE OF A BOILER HOUSE

Pollutant emissions come from stationary sources and motor transport. Boiler plants of various enterprises make a great contribution to pollution. Boiler plants are today an indispensable element of the infrastructure of a modern settlement: they provide heat and hot water to residential and social premises, administrative and household buildings, support production and industrial enterprises.

1.1 Classification of boiler plants

Boiler plant means an engineering structure in a single technical room, which is designed to heat a heat transfer medium for use in heating or heat and steam supply systems.

Fig. 1.1 - Classification of boiler plants by type of fuel used

Figure 1.1 shows that boiler houses are: gas, liquid fuel, solid fuel, and combined multi-fuel. For gas-fuelled boilers natural gas is used. For liquid-fuel boilers, fuel oil, diesel fuel, waste oil are usually used. Wood, brown and hard coal, briquettes are used as solid fuel.

Each type of boiler plant is analysed below by the type of fuel used.

A gas boiler plant is an installation whose main fuel is natural gas. Natural gas has a number of advantages: relatively low price, availability, ease of use and the ability to automate the process of operation. Unlike other fuels, natural gas does not require to install a fuel storage facility and constantly replenish its reserves, to equip a complex supply system (tank and pumps for liquid fuel, hopper and conveyor for solid fuel), manual work (unloading, moving, loading into the boiler furnace) is completely eliminated.

As for the disadvantages that the gas boiler room possesses, the main one is increased danger. The fact is that to ensure complete and rapid combustion of gas, it must be supplied to the gas burner with very high pressure. This requires high precision in the manufacture of fittings and quality connection of pipelines at their joints. The slightest gas leakage can lead to unfavourable consequences, especially if the gas main is located near a permanent source of open fire - the combustion chamber [1-6].

A liquid-fuel boiler plant is a boiler plant whose main fuel is liquid fuel: diesel fuel, oil and oil products, fuel oil, waste oil.

The main advantage of liquid fuel boilers is the autonomy of the heating system, i.e. independence from the gas pipeline. The efficiency of fuel combustion is very high, efficiency up to 94%. In addition, the design of the fuel burner and furnace of the liquid fuel boiler provides automatic operation throughout the heating period, which makes the system easy to maintain and service.

Among the disadvantages is that liquid fuel boilers and burners to them are more expensive than boilers designed for other fuels. It is also necessary to install a special fuel tank. A fine fuel filter is necessary, because the lack of a filter will quickly put the boiler out of operation [7].

A solid fuel boiler plant is a boiler plant that uses the heat of combustion of solid fuel to produce heat, steam or hot water for consumers.

The most pronounced advantage of solid fuel boilers is reliability, because the absence of electronics, mechanics and moving parts make breakdowns almost impossible. Also to the advantages of such boilers can be attributed their "omnivorousness". So, it is possible to burn almost everything that can burn in the boiler. But efficient and stable operation is guaranteed only by the use of quality fuel: coal, wood, fuel briquettes.

Solid fuel boilers also have a number of disadvantages. For example, they are positioned as autonomous, but the whole system requires mandatory circulation when the boiler is operating. Stopping circulation during operation will definitely lead to system failure or even boiler explosion. In addition, in this type of boiler room has the need to install additional equipment for safe operation. Maintenance of the boiler room is of great importance, as solid fuel (especially poor quality) leads to the formation of a large amount of soot, tar, tar on the gas removal system, causing the need to clean the boiler more often and thoroughly [8].

Combined (multi-fuel, universal) boilers have boilers that can change the type of fuel used, as well as use two, three or four types of fuel. Combinations of fuels can be used in many different ways: electricity and liquid fuel, liquid fuel and gas.

This type of boiler has separate combustion chambers for solid, liquid and gaseous fuel, as well as an integrated electric heater. Multi-fuel boilers are able to heat large areas of premises (up to thirty thousand square metres). Also speaking of the advantages of using this type of boiler room, it is worth mentioning the uninterrupted operation of the boiler, as the equipment will function effectively in the absence of any - any type of fuel.

The disadvantages of combined boiler plants include costliness and the need for more complex maintenance [9].

In this paper the environmental aspect is worth paying special attention to, as all types of boiler houses have one significant disadvantage: harmful emissions, products of fuel combustion, which are discharged into the environment. Table 1.1c considers substances emitted from boiler houses using different types of fuel.

Table 1.1 - Comparison of pollutant emissions depending on fuel type

^-^_Boiler type Discardable substances	Gas	Liquid fuel (fuel oil)	Solid fuel (coal)
Oxidazote (N_2O)	+	+	+
Carbon dioxide (CO2)	-	+	+
Nitrogen dioxide (NO)2	+	-	+
Benz(a)pyrene (C20H12)	+	-	+
Sulphur dioxide (SO)2	+	-	+
Inorganic dust	-	-	+

Soot	-	-	+

As can be seen from Table 1.1, when using gaseous fuel, 4 types of pollutants are emitted: nitrogen dioxide, nitrogen oxide, sulphur dioxide and benz(a)pyrene. At the same time, coal combustion in boiler-houses emits almost twice as many substances: nitrogen dioxide, nitrogen oxide, carbon oxide, benz(a)pyrene, sulphur dioxide, inorganic dust and soot. When burning fuel oil, carbon, sulphur and nitrogen oxides are emitted with flue gases.

Harmful chemicals and aerosols also have a detrimental effect on the environment, leading to numerous consequences such as the greenhouse effect and consequently global warming.

1.2 Climate change

Climate change is one of the most discussed topics in the world today. The Special Report on Global Warming prepared by the IPCC (Intergovernmental Panel on Climate Change) and published in October 2018 once again confirms that human activity has the greatest impact on the increase in the average temperature of the planet: "According to some estimates, human activity has caused a temperature increase of about 1.0 °C compared to pre-industrial levels (an approximate range of 0.8 °C to 1.2 °C). At this rate, global warming is likely to reach a temperature increase of 1.5 °C between 2030 and 2052" [10]. [10].

To combat climate change and its negative impacts, countries adopted the Paris Agreement at COP21 in Paris on 12 December 2015. The agreement, which entered into force less than a year later, aims to significantly reduce global greenhouse gas emissions and limit the rise in global temperatures this century.

The Paris Agreement is a legally binding international treaty on climate change. It was adopted by 196 Parties at the 21st session of the Conference of the Parties to the United Nations Framework Convention on Climate Change (COP21 of the UNFCCC) on 12 December 2015 in Paris. The Agreement entered into force on 4 November 2016.

The Paris Agreement aims to keep the increase in global average temperature well below 2 degrees Celsius above pre-industrial levels while endeavouring to limit the temperature rise to 1.5 degrees Celsius.

To fulfil this strategic goal, countries aim to pass the peak of global greenhouse gas emissions as soon as possible in order to build a climate-neutral world by the middle of the 21st century.

The Paris Agreement is a key element of the multilateral climate process. It is the first ever legally binding instrument to unite countries in pursuit of a common goal to combat and adapt to climate change.

In order to achieve the long-term goal of the Paris Agreement, the document calls on countries to develop and submit Long-Term Low Greenhouse Gas Emission Development Strategies to the UN Climate Secretariat by 2020.

The Paris Agreement is a framework to support countries in need of financial, technological and capacity building assistance.

The Paris Agreement stresses that developed countries should take the lead in mobilising financial resources to support more vulnerable States. However, other countries are also encouraged to provide appropriate financial assistance. Climate finance is necessary for climate change mitigation, as large-scale investments are needed to significantly reduce emissions. Climate finance is also important in the context of adaptation, as adapting to a changing climate requires significant financial resources [11].

The Intergovernmental Panel on Climate Change (IPCC) estimates that the average surface temperature of the Earth has increased by 1.0^0 C since 1850. This leads to negative consequences such as:

— changes in the water cycle, leading to intense precipitation and associated flooding in some regions and extreme droughts in others;

— Increased permafrost thawing, loss of snow cover and shrinking glaciers in the Arctic;

— the emergence of more frequent warm tides, which will affect the ocean ecosystem and the people who live off fishing [12];

— increase of the world ocean level by 1-3 metres by the end of the century, which leads to flooding of territories;

— Reduced yields of major cereal crops: a 1°C increase reduces yields by 10 per cent. According to other estimates, the reduction can be as high as 15-17%;

— Everest's "blooming". Climatologists studied satellite photos of the mountain for

1993-2018 and found vegetation at an altitude of 5,500 metres. The slopes, which were previously considered unsuitable for plant and animal life due to extremely low temperatures, have become actively covered with grasses and shrubs in the last 13 years;

— Increased intensity and frequency of abnormal weather events such as hurricanes, floods, droughts and storms;

Reduced catches in the fishing industry and the destruction of coral reefs due to increased acidity in the world's oceans;

— Loss of valuable habitats, biodiversity, important ecosystems and diverse species due to lack of ability to adapt to rapidly changing environmental conditions;

— changes in the geographical habitat of some species.

— rising water levels in the seas and oceans due to the expansion of water at higher temperatures and the melting of glaciers;

— Increased intensity and frequency of abnormal weather events such as hurricanes, floods, droughts and storms;

— desertification of territories and reduction of crop yields due to the depletion of water reserves. This in turn can lead to regional tensions, exacerbating existing conflicts;

— a threat to long-term crop planning and food stocks in general due to less reliable and unpredictable seasonal shifts;

— Reduced catches in the fishing industry and the destruction of coral reefs due to increased acidity in the world's oceans;

— Loss of valuable habitats, biodiversity, important ecosystems and diverse species due to lack of ability to adapt to rapidly changing environmental conditions;

— changes in the geographical habitat of some species;

increase in the spread of diseases such as malaria and Dengue fever, as disease vectors (mosquitoes) will be able to survive over large areas [13].

Over the last century, human activities have also resulted in the emission of greenhouse gases into the atmosphere, thus increasing the natural greenhouse effect. Most of these gases arise from the use of energy, the construction of mining and combustion facilities, and emissions from motor and air transport. The largest share of all greenhouse gases is carbon dioxide (CO_2), which has the greatest impact on global warming.

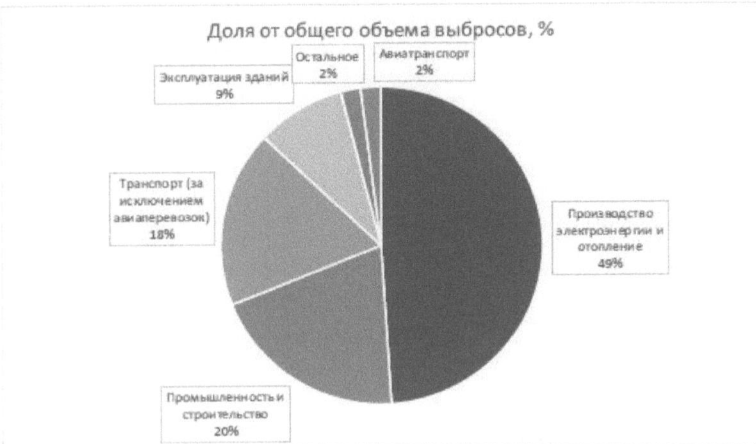

Fig. 1.2 - Sources of CO_2 emissions in the world, share of total emissions, %

As the figure shows, electricity generation and heating account for the bulk of total carbon dioxide emissions, so negative emissions technologies are focused on removing gigatons of carbon dioxide from power generation facilities.

Carbon dioxide (CO_2 , carbon dioxide, carbon dioxide) is a chemical compound without any colour or smell. Normally, carbon dioxide is always present in the air. Its small concentration of 0.03-0.04% is considered safe for humans. The maximum permissible for life activity is 0.15% - 0.2%. However, even a small excess makes carbon dioxide a suffocating gas that can cause serious harm to health [14].

Carbon dioxide is used in the following applications:

— when welding metals to create a protective environment;
— in the food industry for carbonated drinks;
— for cooling, freezing and storing food;
— for fire suppression systems;
— for cleaning various surfaces with dry ice;
— for air guns;
— as an energy source for motors in aeromodelling.

While significant increases in climate action are needed to meet the goals of the Paris Agreement, new low-carbon solutions have been developed over the years and new markets for clean technologies have opened up. More and more countries, regions, cities and companies are setting targets to achieve carbon neutrality.

Carbon neutrality can be achieved through the application of emission reduction or sequestration technologies. There are several technologies available today:

— Reforestation or afforestation;

— Bioenergy with carbon capture and storage;

— increased weathering;

— Carbon capture from the air and storage;

— fertilising the ocean;

— Carbon fixation in the soil.

CHAPTER 2

ASSESSMENT OF GREENHOUSE GAS EMISSIONS AT THE BOILER HOUSE AND DEVELOPMENT OF MEASURES TO REDUCE THEIR ENVIRONMENTAL IMPACT

To assess greenhouse gas emissions at the boiler house of the enterprise under consideration, we will calculate the quantification of pollutant emissions in accordance with the Order of the Ministry of Natural Resources and Environment of Russia dated 30.06.2015 N 300 "On Approval of Methodological Instructions and Guidelines for Quantification of Greenhouse Gas Emissions by Organisations Engaged in Economic and Other Activities in the Russian Federation". Negative emission technologies are also considered and the calculation of a nozzle absorber is performed.

In accordance with the Order of the Ministry of Natural Resources and Environment of the Russian Federation dated 30.06.2015 N 300 "On Approval of Methodological Instructions and Guidelines for Quantitative Determination of Greenhouse Gas Emissions by Organisations Engaged in Economic and Other Activities in the Russian Federation".

In accordance with these Methodological Guidelines, greenhouse gas emissions from stationary sources (boiler house) were calculated.

Fuel consumption in energy equivalent $FC_{j,y}$, t c.e. is determined by the formula:

FC =FC' - k^{N} .n^{v} j,y j,y ,

where

$FC'_{j,y}$ - fuel consumption j in physical terms for the period y, tonnes or thousand m^3 ;

$k_{j,y}$ - conversion coefficient in tonnes of equivalent fuel, t c.t./t, t c.t./ thousand m^3 (based on Table 1.1 of the Order of the Ministry of Natural Resources of Russia dated 30.06.2015 N 300).

$$FC_{j,y} = 851,846 - 0.768 = 654,217.728 \text{ tonnes of fuel equivalent.}$$

CO_2 emission factors from fuel combustion EF y_{CO2y} , t CO_2 /t are determined according to the formula:

EFCO2j,y = WC,j,y - 3.664,

where

$wc_{j,y}$ - carbon content in j-fuel for the period y, t C/t;

3.664 - conversion factor, t CO_2 /T C.

$$EF_{co2,j,y} = 0.75 - 3.664 = 2.748 \text{ t CO2/T}$$

Quantification of CO2 emissions from stationary fuel combustion is performed by calculation method for individual sources, groups of sources or the organisation as a whole according to the formula:

$$E_{CO2,y} = \sum_{j=1}^{n}(FC_{j,y} \cdot EF_{CO2,j,y} \cdot OF_{j,y}),$$

Where:

$E_{CO\,2,y}$ - co2 emissions from stationary combustion of fuel for the period y, t co2;

$FC_{j,y}$ - fuel consumption j for the period y, thousand m^3 , t, t c.t. or TJ;

$EF_{CO\,2,j,y}$ - co2 emission factor from combustion of fuel j for period y, t co2/unit;

$OF_{j,y}$ - oxidation coefficient of fuel j, fraction (taken equal to 1);

j - type of fuel used for combustion;

n - number of fuels used during the period y.

$$E_{CO2,y} = 654{,}217{,}728 - 2{,}748 - 1 = 1{,}797{,}790.32 \text{ tonnes co2e}$$

To reduce the level of negative consequences of its activities, an enterprise can use various technologies to reduce greenhouse gas emissions. The following technologies are currently available:

— Reforestation or afforestation;
— Carbon fixation in the soil;
— Bioenergy with carbon capture and storage;
— increased weathering;
— Carbon capture from the air and storage;
— fertilising the ocean.

In this paper the negative emission technologies given above are studied. Adsorption and absorption methods, reconstruction of a coal-fired boiler house and its conversion to wood waste are also analysed.

2.1 Reforestation or afforestation

Reforestation - growing forests in areas subjected to logging, fires.

One way to mitigate climate change by absorbing the carbon dioxide that is constantly released into the atmosphere from stationary sources and transport is to develop a forest system. Forests and their trees are the main regulator of the water, energy and carbon cycle. In addition, provide the basis for carbon storage, cooling of the earth's surface and water allocation.

In this case, afforestation or reforestation acts as one of the activities that does not emit carbon but absorbs the carbon that is already in the atmosphere.

According to the data of remote sensing of the earth, which were carried out by Russian scientists, the dynamics of the forest area of the Russian Federation is given [15].

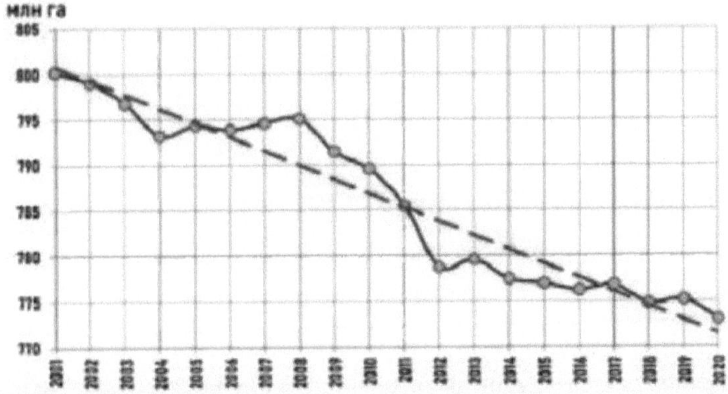

Fig. 2.1 - Dynamics of the Russian forest area in the period from 2001 to 2020.

As shown in Figure 2.1, from 2001 to 2020, the forest area decreased by 27 million ha. The average annual rate of reduction of the forest area of the Russian Federation was 1.4 million ha/year^{-1} . In the periods from 2004 to 2008, an increasing trend in forest area can be seen. A similar trend is seen in the periods from 2012 to 2013, from 2016 to 2017, and from 2018 to 2019, but despite this the forest area continues to decrease.

In order to increase forest area, there is a need to reduce deforestation in addition to cultivation. It is worth noting that carbon storage has increased in temperate areas in recent years, while tropical areas have, conversely, experienced a decrease in forest area due to degradation as well as deforestation. Reducing the rate of tropical deforestation has long been identified as one of the most effective and economic means of limiting emissions.

A pioneering initiative to reduce emissions from deforestation and forest degradation (REDD) is leading the fight against climate change. Forests have the potential to mitigate climate change by more than 5 $GtCO_2$ - eq. per year by halting forest destruction and degradation. A 2020 report by UNEP and the Food and Agriculture

Organisation of the United Nations found that 420 million hectares of forests have been lost to land use expansion over the past 30 years, with another 100 million hectares already threatened with destruction. In the next decade, according to the decision made at the 26th UN Conference of the Parties on Climate Change, all countries of the world will have to stop destroying forests [16]. According to research, forests the size of Portugal are disappearing from the face of the Earth every year, causing climate change and other environmental problems, including forest fires, species extinction and food shortages.

Responsible forest management and successful reforestation strategies are highly relevant at the moment, and one of the FAO (Food and Agriculture Organisation of the United Nations) projects funded by the GCF (Green Climate Fund) and implemented in Argentina aims to do just that. Between 2014 and 2016, the Argentine government managed to prevent the emission of 18.7 million tonnes of carbon dioxide equivalent by successfully reducing deforestation. Building on this successful experience, FAO and GCF are implementing an $82 million project that aims to further reduce deforestation in Argentina [17].

The researchers say that despite attempts to reduce the rate of tropical deforestation, emissions from deforestation continue to increase at a high rate and are accounted for at about 3 billion tonnes of CO2, about 10% of the world's emissions by 2010.

This is primarily due to the decline of tropical forests. Against this background there is a net transfer of carbon from forests and from the forest floor to the atmosphere and researchers have calculated scenarios for future reversal of these trends using extensive reforestation technologies as a means of removing large amounts of CO2 from the atmosphere.

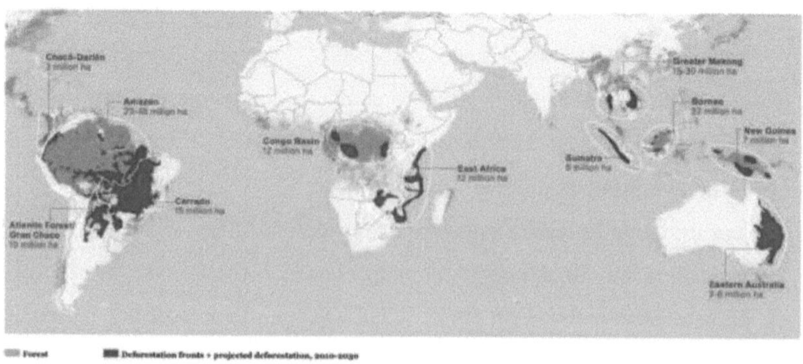

Fig. 2.2 - Map of deforestation

As Figure 2.2 shows, much of the tropical forests have already been cleared, and also large areas of these forests are threatened by deforestation.

Thus, the IPCC has been assessed and the results suggest potential carbon sequestration rates of up to 1.5, either 9.5 or 14 $GtCO_2$ per year. These are 3 different scenarios for human behaviour and carbon dioxide emission and carbon dioxide uptake by forests.

There are two different methods of reforestation - artificial (planting or seeding a forest) and natural without human intervention and with human assistance - promoting natural regeneration (creating conditions for rapid establishment of valuable tree and shrub species).

Table 2.1 summarises the advantages and disadvantages of different methods of afforestation.

Table 2.1 - Advantages and disadvantages of natural and artificial reforestation

	Advantages	Disadvantages
Natural regeneration	Population prospectivity in genetic terms	Slow growth in the first years of life
	Increasing resistance to unfavourable environmental factors	In mixed young stands, frequent thinning is required to regulate the composition
	Multicomponent and complexity of natural plantations formation	The need to apply facilitation measures in some contexts, which increases the costs of
		reforestation
	Ensuring restoration of all useful functions of the forest (water protection, recreational, sanitary and hygienic)	Inability to expand stand composition
	No large monetary and labour costs are required	
Artificial reforestation	Urgency and simultaneity of deforestation	Impoverishment of the natural gene pool of tree

	areas, formation of a given composition and design of forest stands	species
	Lack of competition among trees before crown closure, possibility of initial uniform placement of trees on the area	High cost and complexity of works
	Improvement of forest conditions through tillage and other agro-technical measures	reduced resistance to unfavourable factors
		Early start dates for silvicultural care-
		and their frequency of repetition

As can be seen from Table 2.1, both natural and artificial reforestation methods have their advantages and disadvantages. Considering the need to limit the global temperature increase to 1.5^0 C in a short period of time, it is worth using the artificial reforestation option [18].

According to studies, removal of about 1.1 to 3.3 $GtCO_2$ per year through afforestation of about 320, or 970 million hectares of land on which forest will be planted. It must be realised that it will not be enough to plant a forest, but it is also necessary to provide it with nutrients so that those crops most prone to absorbing carbon will grow most vigorously. The nutrient requirements will also be significant. This study shows that somewhere around 1 mGt of nitrogen and 1 mGt of phosphorus per year will be required for active forest development in newly planted areas.

It should be noted that the impact of forests on the climate is not limited only to the growth of phytomass, as the growth of forests changes the albedo (characteristic of the reflectivity of the surface) of the territory, respectively, more energy will be reflected. Also the cycle of soil carbon balance, hydrological balance changes, hence the temperature in the reserve layer decreases.

In addition, the absence of forests can lead to increased emissions of volatile organic matter (VOCs) from soil as well as microbes from soil, which in turn can form aerosols. These effects are multidirectional, difficult to quantify on a site-by-site basis, and in some cases some trends work to reduce global temperatures while others work to increase global temperatures, so in the absence of an adequate model, afforestation instead of reducing CO2 concentrations may lead to increasing trends that lead to higher temperatures.

Possible problems relate to the release of stored carbon from crop disturbance or land use change, nitrous oxide emissions from increased fertiliser use, and impacts on biodiversity. Nitrous oxide is somewhere around 300 times more active as a greenhouse gas than CO2, so reductions in climate impacts from CO2 removal can be offset by increases in nitrous oxide emissions.

There are also concerns about the availability of land in competition for food with a growing global population and (just like natural terrestrial carbon stocks) the vulnerability of sequestered carbon to harvesting (legal and illegal logging), fire, pests and disease. Carbon stored in living biomass may be unreliable and necessary measures and resources will be needed to protect and maintain expanding forests. Water requirements may be an important constraint, especially in arid regions.

In addition, it must be realised that the reduction in CO2 concentration does not immediately follow the planting of forests. The fact is that the disturbance in the soil due to planting, and consequently the release of CO2 from the soil gives a compensation for the carbon that is absorbed by the growing trees, so it takes several years, or in some cases, several decades for forests to reach their maximum carbon sequestration efficiency. Some researchers say about 20-60 years, depending on the type of forest plantation, location and local conditions.

Figure 2.3 - Carbon dioxide uptake by different tree species

As can be seen from Figure 2.3, spruce has the lowest absorption capacity, so the most rational solution is to plant poplar, as it absorbs 7 times more carbon dioxide than spruce.

An important aspect of the fact that sequestered carbon can be dependent on local political peculiarities. Therefore, in such a long term perspective it is necessary to legislate, including at the international level, the approaches of local communities to those forests that work as a whole for the planet Earth. On 2 July 2021, the Forest Code of the Russian Federation was amended, as a consequence of which several types of monitoring, the concept of "forest nurseries" and information on their establishment, as well as a paragraph concerning reforestation and afforestation appeared. Thanks to the amendments to the legislation of the Russian Federation, our country is following the world trends in environmental conservation, the use of forests as a technology of negative emissions and a natural way to protect the environment from excessive carbon dioxide in the atmosphere.

Thus it can be seen that forests and forest plantations are an important element in the system of global warming containment, in the system of CO2 absorption of the atmosphere, in the system of carbon dioxide pumping into the soil by tree roots, but this requires long time periods, protection from forest fires, regulation of local legislation and prevention of deforestation [18].

2.2 Bioenergy with carbon capture and storage

This technology is included in the list of technologies that have been reviewed by the IPCC.

Bioenergy - energy production from biofuels of various types. Biofuels are fuels made from the products of organisms or organic industrial waste. Bioenergy with carbon sequestration and storage (BECCS) - a hypothetical method of removing CO2 from the atmosphere to mitigate global warming.

Biomass captures carbon dioxide during its growth and stores it as organic material. In the future, energy is generated by burning the fuel derived from the plants. Thus, while generating electricity, the carbon dioxide that is produced by burning biomass is captured by various filters and pumped into the subsurface, thereby effectively removing it from the atmosphere. To affect the climate, the process must be carried out on a global scale.

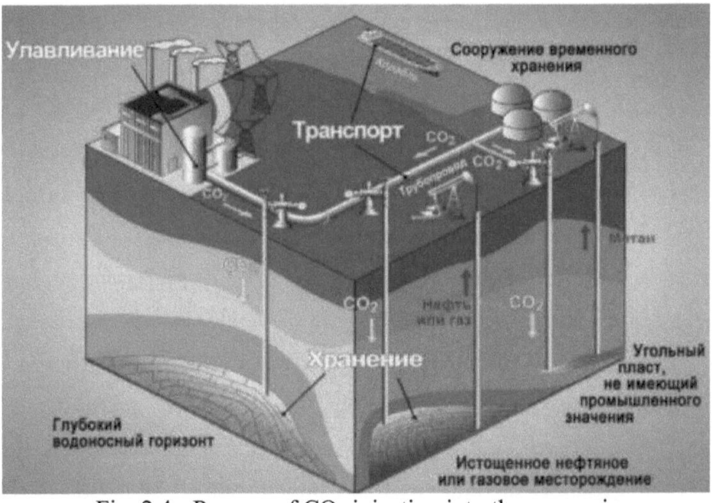

Fig. 2.4 - Process of CO_2 injection into the reservoir

CO_2 is pumped into cavities underground created by pumping out gases, oil, and other natural cavities. The carbon dioxide pumped into them, captured at the end of the pipes during combustion, is stored in these cavities. Later, it enters into chemical reactions with carbonates located underground, and CO2 remains there in the form of various compounds [19].

At the moment, one of the most popular types of biofuel is pellets, produced from peat, wood waste and agricultural waste. They gained their popularity due to the ease of production, easy storage and transportation, safety and cheapness (the average price for 1 tonne of pellets varies from 5 to 10 thousand rubles depending on the season).

The production of wood pellets using forest biomass as feedstock has increased in recent years, and industry consultants (Hawkins Wright, 2019) estimate that global industrial pellet production reached 24 million metric tonnes (Mt) in 2019 (equivalent to a feedstock of ~50 million m^3 wood). The majority of 'industrial' pellet production is for electricity generation, which the International Renewable Energy Agency (IRENA, 2019) categorises as 'solid biofuels and renewable waste'. Here, global generating capacity increased from 52,146 MW in 2009 to 95,687 MW in 2018, with the fastest increases occurring (over the same period) in the EU (from 15,912 to 24,081 MW) and Asia (from 14,140 to 34,845 MW). _Of the pellets produced globally, more than 10 million tonnes are sold internationally, mainly between the United States and the UK and some other European countries, and to South Korea and Japan from Vietnam (lEABioenergy, 2017). This expanding biomass pellet business depends heavily on its relationship to regulations that categorise forest biomass as 'renewable', so many countries have turned to biomass to meet their renewable energy goals.

The classification of forest biomass as "renewable" is based on the reasoning that if biomass carbon comes from atmospheric CO_2 , and forest regrowth absorbs CO_2 over time, it can be considered "carbon neutral" with net emissions during the forest regrowth/deforestation cycle equal to zero. The "carbon neutral" concept, however, is a crude representation of the balance of CO2 in the atmosphere because it ignores the slowness of the photosynthesis process, which takes several decades for trees to grow. This has been repeatedly pointed out by various sources such as Agostini, Guiintoli, & Boulamanti, 2014; Berndes et al., 2016; Fisher, Jackson, & Biewald, 2012; Holtsmark, 2012, 2013. Nevertheless, its simplicity has brought with it political and economic advantages and led to the inclusion of biomass in the European Commission's definition of renewable energy in its 2009 Renewable Energy Directive (RED; EC, 2009), which is seen as "part of the package of measures" needed to reduce greenhouse gas (GHG) emissions.

The RED allowed governments to offer renewable energy subsidies to replace coal in large power plants with biomass, creating a market incentive that led to a rapid increase in demand for forest biomass pellets. However, it is of serious concern that scientific analyses show that instead of reducing greenhouse gas emissions when replacing coal with biomass for electricity generation, CO2 emissions per kWh of electricity will increase as a result of the lower energy density of wood, emissions along the supply chain, and/or less efficient conversion of combustion heat to electricity. The resulting increase in atmospheric CO2 concentration increases radiative forcing and thus contributes to global warming. This initial negative impact is only eliminated later when forested areas grow back. Studies have shown that the time it takes to reabsorb the extra carbon released can be very long, so current policies risk backfiring - initially exacerbating rather than mitigating climate change.

The European Academies Scientific Advisory Council (EASAC) has brought these issues to the attention of the European Commission in its reviews and statements. Also during the discussions on the revision of the EU RED radio equipment directive in 2018, many academics argued against simplistic assumptions about carbon neutrality and treating biomass as renewable. However, the revised directive (REDII) still categorises biomass in the same way as solar, wind and other renewable energy categories. Subsidies continue, and other countries are looking at replacing coal with biomass as a step towards climate change mitigation.

Also of concern is the impact of increased harvesting of forest biomass on ecosystem biodiversity and the loss of processes such as ecological regulation of water and nutrient cycling or soil maintenance.

Forest management has historically included bioenergy production along with construction timber, cardboard production, fibre for paper and other products. Branches, bark and other sawmill residues produced energy. Within the framework of sustainable forest management, this can be seen as maximising the efficient use of available resources, where fuel is produced from materials for which there is no more beneficial use. Such a forest, sustainably managed to maintain a stable or increasing carbon stock, can be characterised as producing no net carbon emissions and thus 'carbon neutral'.

This scenario may have been influential when the EU first identified renewable energy sources to be included in the 2009 RED targets. Prior to this, there was another major decision of the United Nations Framework Convention on Climate Change (UNFCCC) where forestry emissions accounting rules were to be developed under the

Kyoto Protocol. These were based on the assumption that carbon in forests should be treated as released when harvested (regardless of subsequent use). Thus, when it came to accounting for emissions, if forest biomass was utilised, the carbon emitted should (for accounting purposes only) be considered zero, as forestry carbon had already been counted in the category 'land use, land-use change and forestry' (LULUCF). This means that in accounting, emissions from forest biomass used in a power plant are treated as zero, so when the power plant and the country in which it is located report emissions, they are not accounted for. If biomass is collected and combusted in the same country, accounting should reflect overall emission trends in that country. However, an unintended consequence at the time was that the rule allows a country to import biomass, use it for energy production and zero out its emissions, provided they are recorded in the LULUCF statistics of the exporting country. In this way, the importing country can shift the responsibility for reporting its own emissions from forest biomass to the exporting country and be free to dispose of the share of its emissions originating from imported biomass.

The categorisation of biomass as renewable energy has had major implications. Concerns about the intermittent nature of solar and wind power have led governments to seek a 'renewable' supplier of baseload power that can be provided by existing infrastructure. This has led to the replacement of coal with imported wood pellets at a number of plants in the EU (notably in the UK, but also in the Netherlands and Denmark).

EASAC (European Academies' Science Advisory Council), applied the scientific expertise of 27 European academies of science to analyse topical issues related to the interface between science and European policy. One such issue was how best to sustainably manage European forests when they face multiple demands, and EASAC considered the scientific basis for sustainable forest management and the trade-offs between production, biodiversity protection and response to climate change. This and subsequent reports (EASAC, 2018, 2019) addressed the issue of replacing fossil fuels with forest biomass. It was pointed out that:

1. Woody biomass contains less energy than coal (biomass pellets 9.6-12.2 GJ/m3; coal 18.4-23.8 GJ/m3), so CO_2 emissions are higher for the same energy production (110 kg CO_2 /GJ for solid biomass, 94.6-96 kg CO_2 /GJ FOR coal in IPCC, 2006).

Combined with the need for energy harvesting from sources and intermediate processing (drying and pelletising), replacing fossil fuels in electricity generation results in a significant increase in CO2 emissions per kWh. Thus, the net effect of switching to biomass is usually an increase in emissions and therefore an increase in atmospheric CO2 levels. This is the opposite effect to RED's original 'greenhouse gas reduction' targets.

2. Biomass is considered renewable because the CO2 released is assumed to be reabsorbed. However, when forest biomass is burned, carbon from forest stocks is transferred to the atmosphere within minutes, and there is a carbon 'payback period' between this initial release and the return of carbon stocks to the forest through regrowth. This payback period can be several years.

3. An integrated approach is required to assess the net impacts of a transition from coal to forest biomass, where life-cycle carbon fluxes (including emissions from combustion) in a bioenergy scenario are compared to carbon fluxes in the absence of increased production for bioenergy. Such analyses should include the reduction of carbon stocks in deforested forests. Many such studies have shown that only residues from traditional forest management (i.e. residues after use for wood, cardboard, paper, etc.) or naturally rapidly rotting wood from forest die-off from disease or fire have payback periods on the order of a few years. Increased harvesting of stem wood increases atmospheric CO2 levels by decades or centuries, depending on hypothetical scenarios.

4. Even with the shortest payback periods, bioenergy is not the best way to reduce emissions compared to solar and wind, which provide net CO2 savings over months to years. Biomass is thus relatively ineffective for reducing CO2 emissions; yet it is treated equally in regulations and in some EU countries.

5. The sustainability criteria in the RED rules include conditions under which biomass must deliver a certain percentage of greenhouse gas emission reductions compared to fossil fuels. This can easily be misinterpreted to mean that switching from coal to wood is an immediate climate benefit. It is rarely stated that this simply limits supply chain emissions (from cutting, transporting, drying and pelletising, shipping) to less than emissions from burning coal, and ignores the carbon emissions from burning wood.

6. The already mentioned UNFCCC (UN Framework Convention on Climate Change) accounting rules, which allow an importing country to consider emissions from biomass as zero, are based on the assumption that forest biomass reductions are accounted for in the exporting country's LULUCF ("land use, land-use change and forestry") statistics. Since implementation and verification of the latter varies considerably between countries, the trade-off between carbon stock reduction and combustion point emissions is not transparent. Thus, emission reports can be misleading.

The above considerations have led EASAC (the Scientific Advisory Council of the European Academies) to recommend that forest biomass should not be considered as a source of renewable energy under the EU RED unless replacing fossil fuels with biomass results in a net reduction in atmospheric CO2 concentrations within about ten years.

A counterargument to the above is that the removal of carbon storage from a single patch of forest should be considered at a landscape scale, where (at least in some regions) carbon storage may be increasing.

The above describes how the application of the simplistic concept of carbon neutrality has led to costly policies that raise atmospheric CO2 levels and worsen rather than mitigate climate change indefinitely. The Intergovernmental Panel on Climate Change (IPCC) accounting rules, which lump all forestry-related emissions into the LULUCF category, have created a reward for countries importing biomass, although total emissions have likely increased as a result of the shift from coal to imported biomass, a country can account for them as zero and report a decrease. Moreover, energy security, the ability to meet renewable energy targets, and socio-economic benefits in some areas are key aspects that are as important, if not more important to policymakers, than the nuances of the actual impact on climate change. In this situation, it is important to reform current policies to reduce unfavourable climate impacts.

A key component of the new management systems will be to require operators to publish their estimates of the net impact on CO2 levels in the atmosphere throughout the life cycle of their supply chain, including how their feedstock stocks affect current and future carbon stocks. Several methodologies for such calculations are available.

Some debate continues over the sensitivity of such estimates to the choice of reference (counterfactual) scenarios and the role of external factors that may offset the loss of forest carbon stocks, for example, if bioenergy demand leads to improved forest management and productivity, reducing the payback period, or if the IPCC "no bioenergy" scenario leads to natural carbon losses due to pests, fires or other disturbances.

Finally, there is a need for reporting requirements under the UNFCCC that reflect real emissions and their climate impact, and that remove the current perverse incentives to import biomass arising from the

the possibility of treating them as zero emissions at the point of combustion. In the meantime, the EU should reform its own reporting requirements under its emissions trading scheme to ensure that biomass emissions are fully transparent and reflect real climate impacts. Possible alternative reporting criteria have been proposed that would take into account the payback periods of a facility's feedstock in determining the proportion of emissions to be reported, with 'zero emission' limited to facilities that achieve a net reduction in emissions of atmospheric CO2 concentrations on the relevant Paris Agreement timescales.

Reforms such as these would allow the industry to move to methods and scales more compatible with the primary purpose for which it was designed and supported - to achieve net reductions in GHG emissions. -A climate-friendly biomass energy supply chain could continue to provide an additional source of income for foresters by providing demand for low-value intermediate residues from thinning and other wastes, incentivising landowners to maintain their land as forests and keep it healthy and productive. Forestry and its products will continue to play a critical role in climate change mitigation, but from a climate change perspective, optimal use of forests should still maximise their use in construction (lumber and panels), furniture and other products that sequester carbon for human needs over long periods.

In general, bioenergy technology is not limited to industries that generate electricity. Carbon can be integrated into other sectors of the economy, such as hydrogen production, biofuels, biogas.

At the moment there are a few demonstration plants, mainly in the oil refining industry, that separate carbon dioxide and subsequently distill it into rock formations, thus capturing it and preventing CO2 from escaping into the atmosphere.

To achieve negative emissions, biomass combustion must be combined with CO2 capture and storage at the end of the pipe.

A positive aspect of bioenergy with carbon capture and storage is that biomass cultivation is an existing technology, and that carbon capture and storage is also a viable technology. However, spreading this technology to the scale required to remove gigatonnes of carbon would require very large areas of land (land requirements range from ~1 to 1.7 ha for every tonne of carbon equivalent removed each year for forest residues, ~0.6 ha for agricultural residues).

Thus, the technical potential of bioenergy with carbon capture and storage as a negative emission technology needs to be assessed on a case-by-case basis and cannot (as is currently the case in some integrated assessment models) be considered as the primary means of reducing atmospheric CO2 levels [20].

2.3. Carbon fixation in soil

Soils are a major store of carbon. Many cultivated soils have lost 50-70 per cent of their original stored carbon (IPCC, 2000), so farmland management is an important part of soil carbon capture and fixation. Soil carbon content can be increased by growing cover crops, leaving crop residues in the field, applying compost, using low or no-tillage systems and using other land management techniques to stabilise soil structure. The mitigation potential of this approach is estimated to range from 0.07 to 0.7 tCO_2 eq/year per hectare. Soil carbon accumulation can occur for decades before the content reaches a steady state. However, such measures need to be applied continuously as they are easily levelled off if agriculture returns to more intensive tillage practices. Soils contain 1500 GtC at depths up to 1 metre and 2400 Gt at depths up to 2 metres [21].

Changing agricultural practices offer opportunities to increase carbon storage in soils. Scientists have estimated that an increase in soil organic carbon could remove up to 0.7 GtC/year from the atmosphere. Recent studies suggest that the increase in soil organic carbon will peak in 10-20 years when levels approach saturation, but if soil carbon content could be increased in the topsoil, the increased carbon could potentially absorb 2-3 GtC/year.

However, some studies indicate a much more limited potential for removing carbon from the atmosphere on a global scale. Thus, elucidating the potential of land management requires additional research before policy implications can be fully assessed. In the meantime, there are currently no incentives to integrate organic carbon increases for agriculture.

In addition, pyrolysis of biomass to form charcoal (biochar) can produce a soil additive that can sequester carbon in the soil for many years. However, as long as the use of coal and other high-emission fossil fuels is phased out, the alternative use of charcoal as a fuel may have greater potential for climate change mitigation.

2.4. Increased weathering

Enhanced weathering is a technique that characterises geochemical processes that are naturally slow to absorb CO_2 , enhanced by some physical or chemical mechanism.

Figure 2.5 - Types of weathering

Figure 2.5 shows that there are several types of weathering, which are distinguished by the effects of different substances.

This weathering process is believed to stabilise the Earth's climate over geological periods. The use of this technology and its possible acceleration to address the current problem of climate change is currently being raised.

The scientists calculated that for silicate weathering, the theoretical net effect of dissolution and deposition is about 1 mol CO_2 for every mol Ca or Mg of weathered material.

Silicate or carbonate minerals dissolve in rainwater, CO_2 is drawn into solution from the atmosphere. The potential for carbon removal through enhanced weathering (including the addition of carbonate or silicate minerals such as olivine and basalt to oceans and soils) is estimated to be about 1 GtC/year by 2100. One method might involve spreading finely ground silicate mineral rocks over large areas of land, as is already done in some cases to reduce soil acidity in agriculture. Other industrial processes using dilute solutions require large quantities of water. The accelerated weathering concepts that have been explored are the result of theoretical research and limited laboratory testing, and there are no industrial process demonstrations or pilot applications. These methods would require the extraction, transport, and utilisation of very large quantities of minerals (~1 to more than 3 Gt of rock per gigatonne of carbon removed) [20].

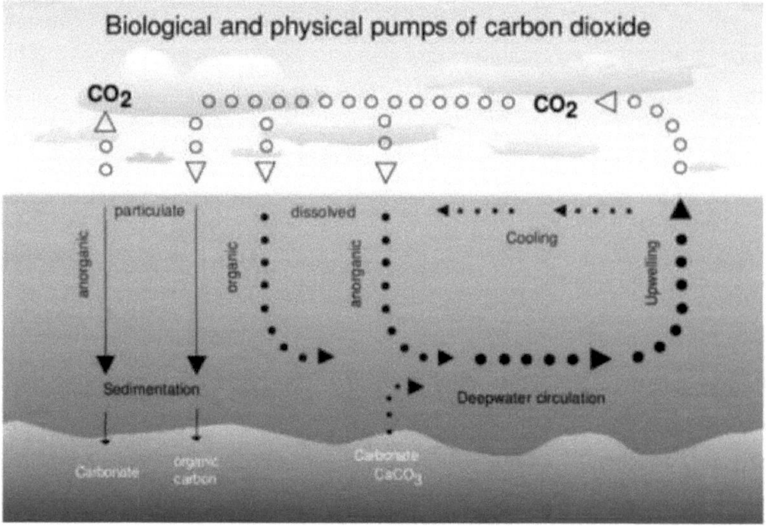

Fig. 2.6 - Scheme of CO mineralisation$_2$

CO_2 dissolves in seawater and slowly mineralises. Eventually, the resulting compound sinks into the deep ocean sediments, but over a rather long period of time, which is measured in hundreds and thousands of years. This process is constant, i.e. deep-sea sediments with carbon dioxide deposited in them are constantly being formed.

The challenge for the researchers is to accelerate the weathering process, coupling silicates and carbonates with CO2, and assessing the potential of this technology.

In this case, the decisive factor is the presence of calcium carbonate or magnesium carbonate, i.e. minerals that enter the ocean after weathering on land, where their alkalinity compensates for the original-

The initial acidification when CO2 dissolves in water. Accelerating this process could increase the rate at which CO2 is taken up by the oceans.

The weathering of silicate minerals and the release of magnesium or calcium silicates into the oceans can affect the chemistry of the processes that occur in water. Similarly, by reacting with carbon dioxide from the atmosphere, biogenic, either dissolved silica or bicarbonate is formed.

It is worth noting that the advantage is the great efficiency of CO2 absorption from the atmosphere, while the disadvantage is the slow reaction rate (tens of thousands of years).

Carbonisation of minerals and acceleration of such natural processes offers us potential means to increase the rate of CO2 removal from the atmosphere on quite large scales, as we are talking about the scale of the areas of the world's oceans.

It is worth noting that minerals react with carbon dioxide to form carbonates through fairly simple chemical reactions, and attempting to recreate weathering on an industrial scale can be quite challenging due to the slow kinetics of these chemical reactions.

Although some researchers say that adding some microbes to aqueous solutions can greatly speed up the process of combining silicates and carbonates with the carbon dioxide found in seawater.

It is also worth remembering that in addition to mining carbonates, it is necessary to transport them for application in those areas or water areas that will be considered and identified by researchers as the most feasible areas or water areas for application of these minerals.

The larger the contact surface area, the more efficient the reaction. In this case inorganic reactions are meant. In order to accelerate the reactions of carbon dioxide absorption from the atmosphere by means of silicates, or calcium and magnesium compounds, it is necessary to give as fine a fraction as possible, and accordingly capacities for crushing minerals are needed. They are then applied to specific areas or water areas.

The researchers estimate that this would require up to 5 kg of silicate rock per 1 m^2 of land area, which would need to be applied to between 15% and 45% of planet Earth's land area (the land area is 149,939,063 km^2) to remove carbon from the atmosphere.

The potential for carbon removal through enhanced weathering by 2100, researchers estimate, could amount to about 1 Gt of net carbon by 2100, for a total of about $50 per tonne of CO_2 .

In addition, industrial wastes, such as various slags, ashes, which are similar in composition to feldspar, for example, and which can be used to sequester carbon from the atmosphere, can be a potential source of minerals. But scientists estimate that removing about 7.5 Mt CO_2 per year from the use of industrial dust products would only reduce carbon dioxide emissions by 0.1 per cent.

To summarise enhanced weathering as a negative emission technology, it is necessary to use a significant amount of energy to crush silicates, transport, deposit them on the territory or water area, and the chemical reaction rate is rather low. In addition, there is no market for large quantities of calcium carbonates or magnesium carbonates. All these are factors that currently limit the application of this technology. Also of great importance are high current costs for large-scale implementation of enhanced weathering technology.

No experimental or demonstration processes are currently in place to determine the scientific, technical and economic advantages and disadvantages, but some field studies have recently begun on the potential of using basalt in agriculture to increase carbon sequestration while increasing agricultural productivity. In general, the land area required to spread and/or dispose of large quantities of crushed silicate or carbonate minerals is large, so that mining and logistics costs can present a major obstacle, compounded by the fact that the likely carbon removal potential is lower than for other technologies [22].

2.5. Ocean fertilisation

Planktonic algae and other microscopic plants absorb CO_2 and convert it into organic matter, some of which sinks as detritus and is isolated in the deep ocean. It is estimated that enhancing this process could significantly affect atmospheric CO_2 concentrations over several decades or centuries. Since photosynthesis processes depend on the availability of nutrients, ocean fertilisation has long been considered as one possible way to capture CO2. The most promising micronutrient investigated to date is iron because of the large carbon to iron ratio in plankton.

However, studies have highlighted several possible drawbacks. Plankton types are not controllable and will depend on other factors (such as the availability of silicon, essential for some species), and iron supplementation may stimulate the growth of diatom species, some of which are associated with toxic algal blooms. Ecological impacts on the marine food web and fisheries, as well as downstream effects on nutrient supply, productivity and food web dynamics, are also difficult to predict. Ideally, ocean fertilisation with iron should promote the formation of plankton with larger cells, which may facilitate faster sinking, but this may affect higher trophic levels including fish, seabirds and marine mammals in unpredictable ways. Moreover, a very large increase in plankton production is required to achieve significant amounts of sequestration, as most of the bound carbon is released by plankton decomposition or consumption, and only a small fraction reaches the ocean depths. Thus, the sequestration potential is unlikely to exceed 1 Gtu per year according to a National Research Council study, and thus ocean iron fertilisation involves a very high level of uncertainty and environmental risks with a relatively small sequestration potential [20].

2.6. Carbon Capture and Storage from Air (DACCS)

Direct CO_2 air capture involves a system in which air from the atmosphere passes through a contactor/closing device that removes CO_2 , which is then released as a concentrated stream for removal or use, while the sorbent is regenerated and the CO2-depleted air is returned to the atmosphere. Key potential technologies include liquid or solid absorbents.

On the positive side, the removal with liquid absorbents (e.g. monoethanolamine) of more concentrated CO2 from natural gas is already in use, and is in the demonstration phase for the removal of CO2 from flue gases from power plants or industrial plants. The first commercial direct air capture plant (990 tonnes of CO2 per year) was commissioned in 2017 in Switzerland using a solid-state adsorption process. Thus, viable technologies exist today and are at the development or commercial prototype stage for various capture processes. The requirements are fast and efficient absorption (by absorption or adsorption), easy CO_2 extraction and sorbent regeneration mechanisms with low energy costs for air treatment and CO extraction$_2$, as well as long-term storage of the collected CO2.

The negatives include the size and capital cost of the equipment due to the very large amount of air that must be passed through the contactor, the energy losses associated with sorbent regeneration, as well as the need for water to replace evaporation and the potential impact of low CO2 on nearby vegetation.

For the time being, creative ways to minimise energy requirements continue to emerge (e.g. local wind, solar) and new sorbents continue to offer process alternatives that can significantly reduce costs. A recent study concluded that even with large-scale deployment, hypothetical and successful bioenergy with carbon capture and storage would not be sufficient to meet the goals of the Paris Agreement, but substantial deployment of carbon capture from the air and storage (removing several gigatons of carbon per year by 2100) would achieve these goals [23].

2.7. Conversion of coal-fired boiler plant to wood waste boiler plant

A sawdust boiler plant is a solid fuel boiler plant, which uses wood waste as fuel for heat energy generation: shavings, sawdust, crushed bark, wood chips and other materials. Sawdust boilers are subdivided into steam and hot-water boilers. They are in demand at woodworking enterprises, forestry enterprises, for utilisation of woodworking waste and simultaneous heat or steam production, for heating and hot water supply systems, technological steam supply, drying chambers.

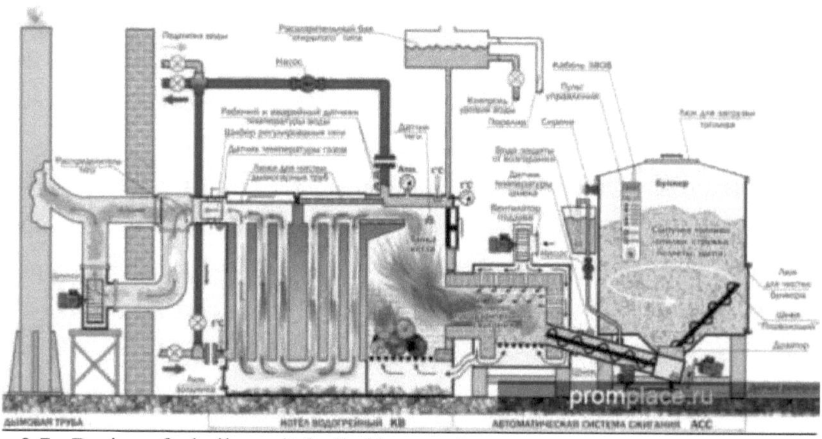

Fig. 2.7 - Design of a boiler unit fuelled by wood waste

Sawdust-fired boilers differ from traditional solid fuel-fired boilers by the presence of remote gas-generator unit and fuel hopper, which is presented in Figure 2.7.

Steam or hot water boilers with sawdust gas generators consist of:

1. a hot water or steam boiler (it heats water or produces and supplies steam using energy obtained from the combustion of gas produced in a gas generator);
2. gas generator (it converts wood waste into combustible gas);

3. fuel hopper with a tedder (it feeds the bulk waste into the gas generator by means of a speckled metering unit);
4. safety and control systems of the boiler and gas generator set.

At the same time, the design of the industrial coal-fired boiler consists of the following elements:

- Smoke extraction system - there is an extractor fan in the design to ensure complete removal of smoke residues from the firebox.

Forced smoke extraction prevents smoke from entering the boiler room and improves draught characteristics.

- Electric ignition - ignition of the industrial coal-fired boiler is carried out by means of electric arc and other methods. The most effective is the method of plasma ignition of dust. In the furnace chamber there is an optical sensor that monitors the presence of flame. When the flame dies out, a signal is given for reignition.
- Loading coal into the boiler - depending on the model, loading is done automatically (by means of a screw gear) or manually.
- Circuit - the design provides for a heating circuit. DHW is provided by connecting an indirect boiler.
- Automation - the operation of automatic and semi-automatic boilers (manual loading) is fully controlled by sensitive automation. Controllers provide the necessary intensity of air supply, monitor the presence of flame, heating medium. The design provides a safety system that prevents boiling of the coolant. Modern models are connected to GSM-modules, allowing to control the boiler remotely [24].

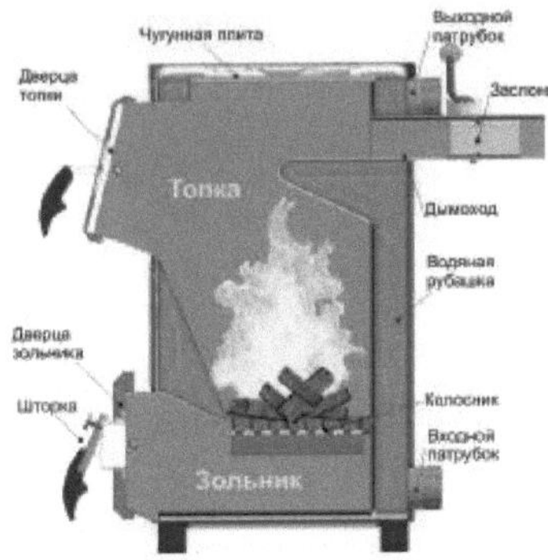

Fig. 2.8 - Coal-fired boiler house

Table 2.2 - Comparison of coal-fired and wood-fired boiler houses

wastes

	Coal	Wood waste
Advantages	Availability and low cost of coal fuel	Cost-effective (especially in the wood processing industry)
	Autonomy (independence from main energy resources - gas pipeline water;)	Automation of work
	Economic efficiency, efficiency of coal-fired boiler house is 84 %	The power of the gas generator can be controlled by adjusting it according to the temperature of the heating medium in the boiler.
	Uninterrupted operation in harsh climatic conditions	Reliability in operation - mechanised boiler houses in autonomous mode are able to work the whole heating season without failures.
		Waste disposal
Disadvantages	Coal combustion is very polluting	Utilisation of waste wood with a maximum *moisture content of 20%.*
	coal storage requires considerable space;	The high cost of a mechanised boiler room.
	Daily maintenance of fireboxes is required;	Waste from the wood processing industry is heterogeneous in terms of density, appearance and fractional composition. Boiler equipment must therefore be versatile
	the combustion process must be monitored at all times.	high temperature of boiler internal walls

Table 2.2 summarises the main advantages and disadvantages of the boilers

The table shows that there are both advantages and disadvantages. The table shows that there are both pluses,

There are enough disadvantages for both the coal-fired boiler plant and the boiler plant operating on wood waste. It should be taken into account that the installation of boiler equipment in this work takes place at the object of wood processing industry, which is an undeniable advantage, as wood waste will be in abundance [24].

One of the ways to improve the efficiency of energy resources utilisation is to increase the share of renewable energy sources in the region's energy balance. This is especially relevant for the Urals and West Siberian regions, where the timber and wood processing industries are developed, and, consequently, the volumes of wood waste - biological fuel for boiler houses - are high.

The economic and environmental benefits of introducing energy equipment using wood waste as fuel at timber processing and woodworking enterprises are obvious: reduction of fuel costs at the constant increase in gas and liquid fuel prices; solving the issues of wood waste utilisation; reduction of harmful emissions.

In the second half of 2005, EcoTherm specialists were asked about the possibility of converting two coal-fired boilers to wood waste in Bodaibo (one of the northern cities of Irkutsk Oblast).

Initially, the economic situation in the region related to the cost of different fuels was analysed. From the results of the analysis it follows that there are significant financial overruns when using diesel fuel and some savings when using wood waste boilers compared to coal-fired boilers.

However, apart from the price of fuel directly, there are other cost items to consider when choosing boilers, namely:

— fuel delivery costs (renewal of fuel stock);

— cost of plant maintenance (frequency and complexity of maintenance, in particular for low-grade coal boilers);

— the increase in the price of fuel over time;

— the cost of installing additional equipment for boiler operation (complex system of fuel preparation and storage).

On-site inspection of the facility revealed the following situation. An existing boiler house for heating production and storage facilities, with a heat capacity of 0.5 Gcal/h with two hot water boilers, uses imported coal as fuel. There is no fuel storehouse and therefore coal is stored in bulk in the open air. There are no fuel supply and ash removal systems on the boilers. It is not possible to speak about good operating conditions of the boiler house.

Coal is delivered to this region only during the summer period during river navigation. There are no other transport interchanges. Therefore, the cost of coal including delivery is significant. Given the harsh climatic conditions in the region, fuel storage is costly.

On the other hand, the enterprise has an existing wood processing facility, which has accumulated a significant amount of sawdust, shavings, chips and lump waste. At the same time, there are no facilities for drying sawn timber, which affects the quality of manufactured products. After the survey, the terms of reference were defined as follows:

1. To propose a technical solution for reconstruction of the existing boilers with the possibility of their operation on wood waste, i.e. to completely exclude coal as fuel;
2. Provide at least a three-day supply of fuel (wood waste) and fully mechanise the process of fuel supply to the boilers;
3. To create a reserve of heat capacity of the boiler house to ensure heating load, as well as heating of the lumber drying chambers;
4. Minimise the distance from the boiler house to future drying chambers to avoid heat loss and to install drying chamber control cabinets in the boiler house itself so that the boiler house operator can control the drying process;
5. Select modularisation of the main boiler and auxiliary equipment to ensure minimum time and ease of installation.

Table 2.3 - Comparison of different fuels

Type of fuel	Calorific value	Price
Diesel fuel	11.9 kWh	59.8 RUB/litre
Coal	7.5 kWh	12.5 RUB/kg
Wood waste	2.3 kWh	6.41 RUB/kg

Table 2.3 presents the calorific values for liquid fuels as well as for the two solid fuels. The table shows that diesel fuel has the highest calorific value, which is almost 5 times higher than the calorific value of wood waste. At the same time, the price per litre of diesel fuel is almost 9 times higher.

In order to reduce reconstruction costs, it was proposed to keep the existing boilers with minor technical changes and use them as heat exchangers. A gas-generator furnace was used for wood waste combustion, which allows burning waste of high moisture content and different composition (from small sawdust to bark). The gas-generator furnace and the heat exchanger are connected by a flame tube, where the final complete combustion of the generator gas takes place, from which the hot gases with a temperature of 1100^{0} C enter the heat exchanger. The simplified scheme of the reconstructed boiler unit is shown in Fig. 2.9.

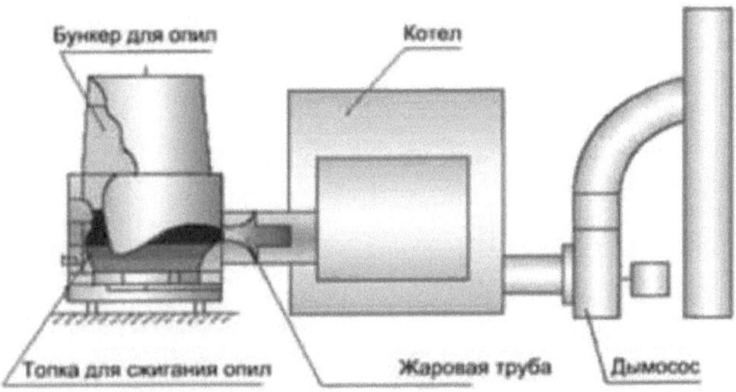

Fig. 2.9 - Simplified scheme of the reconstructed boiler unit

Minor modifications were made to the existing coal-fired boilers, which were used as heat exchangers: an explosion valve was installed; the grate was removed; a smoke pump with the necessary head to ensure normal rarefaction was selected and installed.

As a result of the renovation:

1. The enterprise has fully provided itself with cheap heat energy for heating warehouses, household and production facilities, as well as for technological installations (drying chambers);
2. The cost of delivering coal for the boiler house to this hard-to-reach region has been eliminated;
3. The problem of waste removal from woodworking production has been solved;
4. The operating conditions of the boiler house have improved, the manual labour of the personnel has been reduced. Such a labour-intensive process as bottom ash removal was eliminated, as wood waste has very low ash content.

Currently, the problem of efficient use of energy resources is most acute for budget organisations and public utilities. Rising prices for traditional fuels have an extremely negative impact on the results of financial and economic activities of heat supplying organisations of housing and communal services, on the deterioration of the condition of the main and auxiliary heat and power equipment and networks. This article presents one of the real practical examples of implementation of a rational and economically favourable technical project to solve the accumulated problems.

2.8. Absorption and adsorption

Methods of purification of gases from carbon dioxide are divided into the following groups:

— physical absorption;

— chemisorption;

— adsorption;

— catalytic hydrogenation;

— membrane application;

— enzyme application.

2.8.1. Physical absorption

Physical absorption is based on the solubility of carbon dioxide in polar solvents (water, methanol).

2.8.1.1. Absorption by water

Absorption of carbon dioxide by water is of industrial importance for purification of some high-pressure gases (e.g. for ammonia synthesis). The main advantages of water as an absorbent for removing impurities from gas are its availability and low cost. The use of any other absorbent involves the need for a sealed system and recovery, as the absorbent volatilises during the purification process and the off-gases pollute the atmosphere. Water can be used with less risk of gas leakage and in most cases without recirculation with discharge of saturated solution. To increase the solubility of carbon dioxide in water, the process is carried out at elevated pressure (1.5-2.5 MPa).

The scheme of the installation of gas purification from carbon dioxide by water absorption is shown in Fig. 2.10.

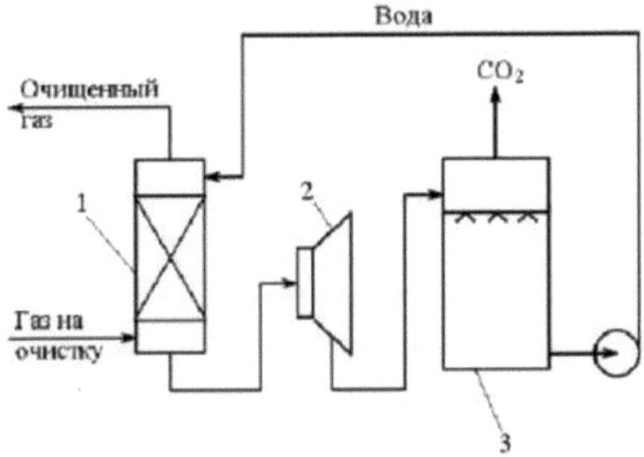

Fig. 2.10 - Schematic diagram of the installation of gas purification from carbon dioxide by water absorption: 1 - absorber; 2 - recuperation turbine; 3 - desorber

The purified gas enters the absorber, where it is flushed with water at a temperature of 30-40° C. The filled absorbent is directed to the desorber, passing through a recuperation turbine, allowing to utilise part of the energy by reducing the liquid pressure and subsequent expansion of the absorbed gas. In the desorber, carbon dioxide is released from the water due to the reduced pressure. The regenerated absorbent is pumped back to the absorber.

The advantage of the process of water absorption is the simplicity of the plant design. Disadvantages: low absorption capacity of carbon dioxide by water (8 kg CO_2 per 100 kg of absorbent); insufficient purity of the released carbon dioxide - 85% (the remaining 15% - hydrogen sulphide, carbon monoxide, nitrogen and other gases).

2.8.1.2. Methanol absorption

Methanol is a good absorber of carbon dioxide. Low temperature and increasing pressure in the system contribute to a sharp increase in the absorption capacity of methanol. At temperature minus 60° C and atmospheric pressure the solubility is 75 cm^3 /g, with pressure increase up to 0.4 MPa it reaches 600 cm^3 /g. Dissolution proceeds with a significant exothermic effect.

Purification of the gas (71.5% H2; 0.8 ¹>, 0.9% CH4; 2.4% CO; 24.4% CO2) obtained by carbon monoxide conversion produces a gas of composition: 92.8% N2; 1.1% CN4; 3.1% CO; 2.0% CO2.

Absorption includes:

2.8.1.3. cooling the gas to minus 35 °C by direct contact with a refrigerant - 60% aqueous methanol solution;

2.8.1.4. absorption proper in the column at temperature minus 35 - minus 60° C and pressure 2 MPa;

2.8.1.5. desorption by successively reducing the pressure to 800, 130, 60 and 25 kPa.

Complete regeneration of methanol is achieved by boiling it.

The advantage of the process - solubility of CO2 in methanol is much higher than in water. Disadvantage - bulky equipment.

Besides methanol, other organic solvents can be used: K-methyl-pyrrolidone, sulfolane, propylene carbonate, the price of which is much higher. At the moment they are not widely used.

2.8.2. Chemisorption methods

Chemisorption is based on the chemical interaction of carbon dioxide with compounds of alkaline nature (alkali, ethanol mines, carbonate solutions, calcium oxide).

2.8.2.1. Purification of gases with aqueous monoethanolamine solution

Chemisorption of carbon dioxide with ethanolamines is used in the preparation of various process gases for processing.

Monoethanolamine (MEA) has the maximum absorption capacity towards carbon dioxide.

In the process of carbon dioxide chemisorption 15 - 20% solutions of monoethanolamine are usually used. Chemisorption proceeds at temperature 40 - 45° C and pressure 1.5 - 3.0 MPa. The resulting carbonates and bicarbonates are decomposed in the desorber with carbon dioxide release when the flow is heated to 120° C.

The scheme of the monoethanolamine gas purification unit is shown in Fig. 2.11.

Raw gas is fed to the bottom of absorber 2 and acid gas absorber to the top. Coming out of the absorber top, the purified gas enters the separator 1 for separation of carried away droplets of MEA solution. The saturated absorber solution enters the vessel 3 for separation from dissolved hydrocarbon gases, then passes through the heat exchanger 6 and is directed to the evaporation column 7. Sour gases, water vapour and MEA from the top of evaporation column 7 after cooling and condensation in condenser-cooler 9 go to separator 10, where carbon dioxide is separated from condensate, which is pumped by pump 11 for column irrigation. Regenerated solution from the bottom of the column passes through the heat exchanger 6, cooler 5 and enters the tank 4, from where it is directed to the absorber 2 irrigation.

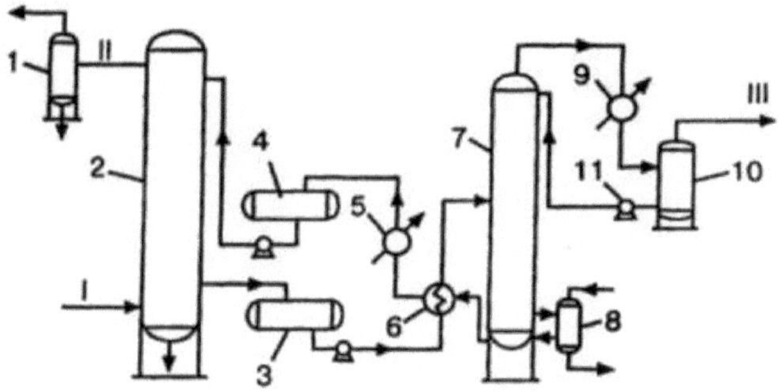

Fig. 2.11 - Scheme of the installation for gas purification from CO_2 by monoethanolamine solution I - raw gas; II - purified gas; III - carbon dioxide; 1, 10 - separators; 2 - absorber; 3, 4 - tanks; 5 - refrigerator; 6 - heat exchanger; 7 - evaporation column; 8 - boiling boiler; 9 - condenser - refrigerator; 11 - pump.

Disadvantages: significant heat consumption for sorbent regeneration, increasing with increasing concentration of carbon dioxide in purified gas; losses of relatively volatile absorbent; formation of significant amounts of resinous substances due to cyclisation of monoethanolamine into oxazolidone-2 derivatives and further into their transformation products.

2.8.2.2. Cleaning with carbonate solutions

This method is based on the interaction of carbon dioxide with aqueous solutions of sodium and potassium carbonates (usually potash) with activating additives of polyvalent metal oxides. To increase the solubility of carbon dioxide in chemisorbent and the speed of the absorption process, hot solutions of carbonates (110 - 120 °C) are used. In practice, 25% aqueous solution of potash activated with arsenic is used. Pressure at the absorption stage is 1 - 2 MPa.

Desorption gases are discharged from the regenerator in two streams in the form of "clean" and "dirty" fractions. "Clean" fraction, constituting about 85% of the total CO_2 , is withdrawn below the point of saturated solution introduction into the regenerator with temperature not exceeding 100° C and with overpressure higher than 0.039 MPa. It is cooled down in the air cooler 2 to the temperature below 57 °C and after moisture separation in the separator 3 is supplied through the electric gate valve 6 to the consumer. The content of combustible components in the "clean" flow does not exceed 0.2% vol. After passing through separator 4, the "dirty" fraction is sent to the liquid carbon dioxide production unit.

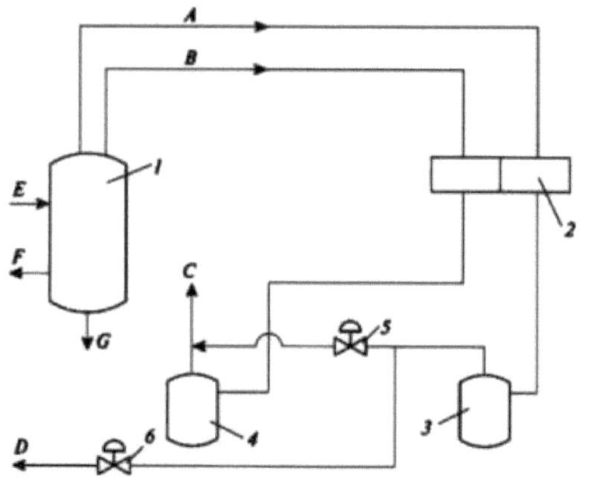

Fig. 2.12 - Principal technological scheme of regeneration of saturated solution of potash: 1 - regenerator (desorber); 2 - air cooler ; 3, 4 - separators; 5, 6 - gate valves; A - "clean" CO_2 ; B - "dirty " CO_2 ; C - "dirty" CO_2 to processing; D - "clean" CO2 to consumer; E - input of carbon dioxide saturated solution; F, G - outputs of " poor" and "half-poor" solutions.

The degree of gas purification from carbon dioxide by potash solution is somewhat lower than in case of purification by monoethanolamine solution. The residual content of carbon dioxide in the purified mixture after the use of potash is 0.05-0.1 % May. However, compared to monoethanolamine, potash is a chemisorbent chemisorbent cheaper.

Well-proven in industry method of process gas purification from CO_2 is used in ammonia synthesis units of large capacity. The purification scheme (process "Karsol") is characterised by good technical and economic indicators (low heat consumption, low costs for heat exchange surfaces). Absorbent of composition 25 - 28 % K2CO3 -

absorbent, 1.9 % diethanolamine - activator, 0.4 % U205 - corrosion inhibitor, process temperature 100 - 103° C, pressure 1.9 - 2.7 mPa are used.

2.8.3. Adsorption

Adsorption is based on the absorption of carbon dioxide by solid sorbents (zeolites).

2.8.3.1 Adsorption by zeolites

Zeolites are efficient adsorbents of carbon dioxide. The diameter of carbon dioxide molecules is about 0.31 nm, which allows them to penetrate into the internal structure of most zeolites. The most common zeolite used in adsorption is CaA.

Fig. 2.13 - Synthetic zeolite CaA - CO absorber$_2$, aluminosili cat with high degree of resistance to weakly acidic media

Figure 5.13 shows zeolite_ in pellets ranging in size from 3 to 5 mm and 1.6 to 2.5 mm. Together with carbon dioxide zeolites absorb water vapours. Therefore, simultaneously with purification of gases from carbon dioxide their drying takes place. Desorption of absorbed components is carried out by pressure decrease and temperature increase. The scheme of adsorption purification of gas from carbon dioxide by zeolite is shown in Fig. 2.14.

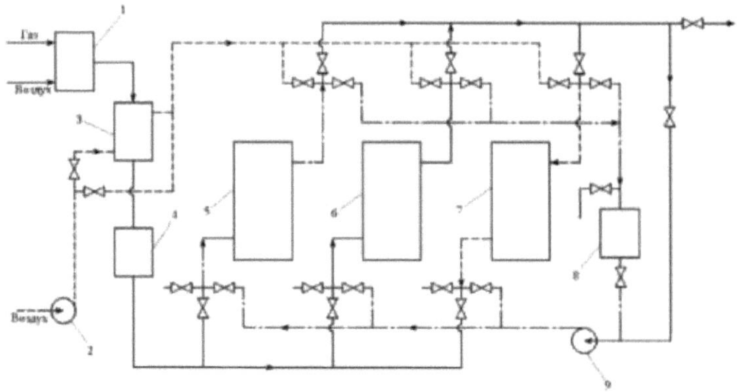

Fig. 2.14 - Scheme of adsorption purification of gas from carbon dioxide by zeolite CaA 17: 1 - gas generator; 2 - fan; 3 - gas-air heat exchanger; 4, 8 - water coolers; 5 - 7 - adsorbers; 9 - circulating gas blower.

Natural gas is combusted with air oxygen taken in deficiency in gas generator 1. Products of incomplete combustion containing $C0_2$, after cooling in heat exchangers 3 and 4 and moisture extraction go to one of adsorbers 5, 6 or 7 filled with zeolite CaA, where gas drying and its purification from C02 takes place. Regeneration of zeolite is carried out by air heated in heat exchanger 3 and supplied by fan 2. Zeolites at the stage of regeneration are heated to the temperature not lower than 235° C. They are cooled by pure gas circulating through the system and cooler 8 with the help of gas blower 9. In the initial period of cooling the gas after blowing through zeolites is discharged into the atmosphere, and then circulates in a closed circuit with constant feeding with fresh gas.

2.8.4. Catalytic hydrogenation

Catalytic hydrogenation is used to remove small amounts (on the order of fractions of a per cent in a mixture of gases) of carbon dioxide. The method is based on the following reaction: $CO_2 + 4N_2$ CN_4 + 2H2O.

The reactions take place at pressures of 0.1 - 30 MPa, temperatures of 300 - 350° C using an iron-based catalyst or at 200° C using a nickel-chromium or nickel-aluminium catalyst.

The catalysts active in the CO hydrogenation reaction are usually also active in the

CO2 hydrogenation. The largest number of studies aimed at the development of catalysts for CO2 methanation is devoted to nickel catalysts. A wide range of carriers is used for synthesis: aluminium oxides, aluminosilicates, diatomaceous earth, ceramics, kaolin, calcium aluminates, etc.

AM type catalysts are produced by the method of deposition on a carrier. To the solution of nickel nitrate with careful mixing add finely ground carrier, for example, aluminium hydroxide (technical), alumina of GA-8 grade, active aluminium oxide of A-1 grade, then soda solution. The best results are obtained when using active aluminium oxide.

Fig. 2.15 - Geometric shapes of methanation catalyst of NIAP-07-07 (NKM-7) grade

Novomoskovsk Institute has developed nickel methanation catalysts by mixing, heterogeneous ion exchange and chemical mixing methods (NKM-1, NKM-2, NKM-3, NKM-4, NKM-4A, NKM-5, NKM-7, NG, etc.).

At present it is known that the most active catalysts for CO methanation$_2$ are group VIII metals: ruthenium, rhodium, platinum, palladium, nickel. However, nickel is prone to carbonisation, and the use of noble metals due to their high cost requires reduction of their consumption. Applied catalysts based on metal complexes of ruthenium and rhodium with trioctylamine (TOA) composition have been proposed. To recycle carbon dioxide into methane, a recognised industrial method involves the reaction of hydrogen and carbon dioxide using a ruthenium-based catalyst at temperatures between 400 and 500 °C.

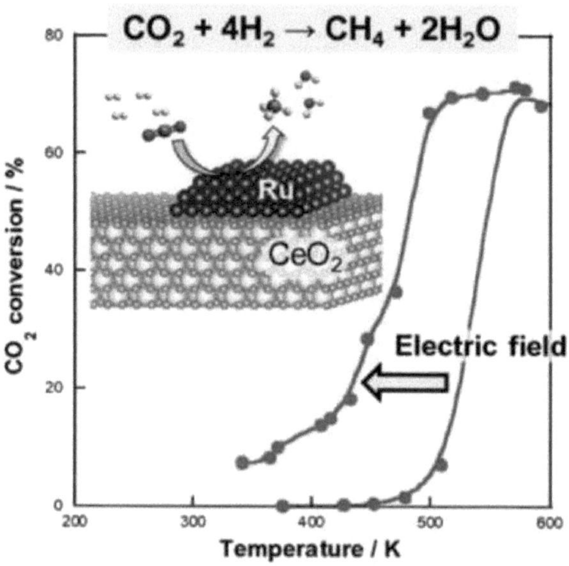

Fig. 2.16 - Methanation of CO_2 on Ru-Ce catalyst under applied electric field (red line) and without applied electric field (blue line)

Figure 2.16 shows a new method of CO methanation$_2$, which involves the reaction of cerium oxide nanoparticles with carbon dioxide in the presence of a ruthenium catalyst under the application of an electric field. The results show that in the presence of electric field, the catalyst exhibits high and stable catalytic activity in the hydrogenation of carbon dioxide to methane with higher conversion than without electric influence.

2.8.5. Membrane extraction

The process of membrane separation of gas components is based on their different ability to pass through a semipermeable partition separating the mass transfer apparatus into two working zones.

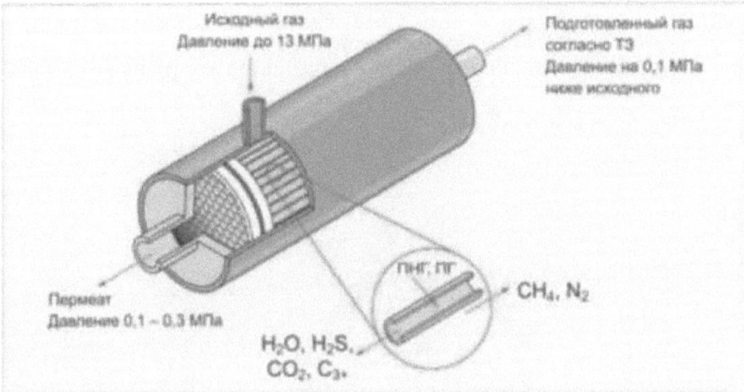

Fig. 2.17 - Scheme of gas mixture separation by membrane method

Figure 2.17 shows how under pressure the separated gas mixture is fed into the pressure drop, easily permeable components pass through the membrane, difficult permeable components are removed from the separation apparatus. Natural rubber, polyvinyltrimethylsilane, Teflon and others, as well as block copolymers of highly permeable siloxanes with monomers such as acrylate are used as the polymeric part of the asymmetric membrane

Lats.

The most important are asymmetric (or anisotropic) membranes, which are characterised by high performance and selectivity. They consist of a thin (0.1 - 1 μm thick, and in some cases 0.005 μm thick) non-porous polymer material applied to a microporous substrate (its thickness is 10 - 100 μm or more).

2.8.6. Enzymatic extraction

One of the most promising ways of CO2 extraction from a mixture of gases is the use of enzymes - organic substances of protein nature.

A key role in enzymatic CO2 capture is played by carboanhydrase, which mimics a natural human lung enzyme that captures and removes CO2 from the blood and tissues. It catalyses a chemical reaction between carbon dioxide and water, converting carbon dioxide into bicarbonate, which can then be recycled to produce baking soda and chalk.

To operate under industrial conditions, the enzyme is immobilised with a solvent inside a reactor. When the flue gas passes through the solvent, the enzyme converts carbon dioxide into bicarbonate.

It is expected that the catalyst forming a monomolecular layer at the water-liquid interface will accelerate the reaction of CO_2 with water to form bicarbonate ion and proton. This approach is promising for CO_2 capture in power plants.

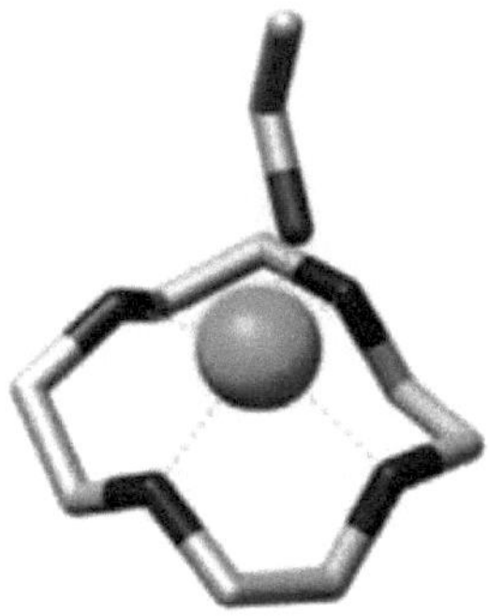

Figure 2.18 - Molecule of zinc complex (green) with ligand 1,4,7,10-tetraazacyclododecane and ethanol

In 2012, a catalyst mimicking the enzyme carboanhydrase was developed (Roger Ains, USA). The proposed molecule is a cyclen with a zinc atom inside and in complex with ethanol, shown in Figure 2.18. Such a complex behaves similarly to the enzyme carboanhydrase and is able to capture a CO_2 molecule at high temperatures, occupying the spatial position of ethanol.

The advantages of enzymatic extraction are environmental friendliness and cost-effectiveness. The apparatus in which absorption processes are carried out are called absorbers. Like other mass transfer processes, absorption takes place at the interface. Therefore, absorbers must have a developed contact surface between the liquid and the gas. According to the method of formation of this surface, absorbers can be conditionally divided into the following groups: surface and film absorbers; nozzle absorbers; barbotage (plate) absorbers; atomising absorbers.

Nozzle absorbers are the most widely used in industry. They are columns filled with nozzles - solids of various shapes. Plate absorbers are vertical columns inside which horizontal partitions - plates - are placed at a certain distance from each other. With the help of plates the directed movement of phases and multiple interaction of liquid and gas is realised.

The purpose of the calculation is to determine the amount of absorbed gas, absorbent flow rate, absorber diameter, column height, hydraulic resistance, and fan power.

To calculate absorbers, the gas flow rate, the gas composition, the initial concentration of the extracted component in the gas phase are specified. In addition, the final concentration of the extracted component in the gas or the recovery rate must be known.

The type of absorbent and its initial concentration are specified. Often the absorbent excess ratio is specified, which is usually taken in the range of 1.1 to 1.3 (i.e. the absorbent flow rate is 10 - 30 % higher than the minimum flow rate). The main quantities to be determined are: absorbent flow rate, absorber diameter and height, absorber hydraulic resistance.

In order to solve the problem of high concentration of aerosols of mainly fibrogenic action at the considered boiler house, the calculation of absorption plants with a nozzle type of column was carried out.

2.8.7. Calculation of the nozzle absorber

To reduce carbon dioxide emissions at the enterprise under consideration, it is proposed to calculate the indicators required for the installation of a nozzle absorber. Calculation is carried out according to the methodological manual Lekanova T.L., Kazakova E.G. "Calculation of absorption plants". Initial data are taken from Table 2.4.

Table 2.4 - Initial data

V, m /s^3	at$_н$, %	Uk, %	г	P, kPa	t, °C	%H
1,05	6	0,4	1,2	141,5	35	0

where

V - quantity of processed mixture, m^3 /s; yy - initial concentration of CO_2 in the mixture, %;

Uk - final concentration of CO_2 in the mixture, %;

£ - absorbent surplus

P - absorption pressure, kPa;

t - absorption temperature, C;

$x_н$ is the initial concentration of CO_2 in the absorbent.

It must be determined:

— quantity of absorbed gas Gco2, kg/s;

— absorbent flow rate L, kg/s;

— absorber diameter D_K , m;

— column height $H_к$, m;

— hydraulic resistance AP, Pa;

— fan power N_B , kW.

The installation scheme is shown in Figure 2.19.

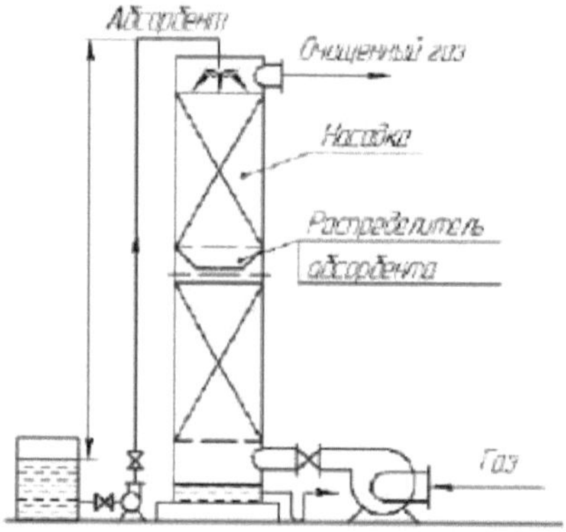

Fig. 2.19 - Schematic diagram of a nozzle absorber for water absorption of carbon dioxide CO_2 from its mixture with air 1.

Determining the amount of carbon dioxide absorbed The amount of carbon dioxide (CO2) absorbed is determined using the formula:

$$G = {}_{CO2}\frac{V \cdot y_{н} \cdot M_{CO2} \cdot \alpha}{22,4}, \text{ kg/s} \quad (2.1)$$

where

$y_н$ is the initial concentration of CO_2 in the gas mixture expressed in mole fractions, $\frac{кмоль\ CO_2}{кмоль\ смеси}$;

M_{CO2} is the molar mass of CO_2 , $\frac{кг}{кмоль}$;

α - CO recovery factor$_2$;

22.4 is the volume occupied by 1 kmol of gas, m^3 .

The initial concentration of CO_2 in the gas mixture is determined by the formula:

$$y_н = \frac{\frac{\bar{y}_н}{M_{CO2}}}{\frac{\bar{y}_н}{M_{CO2}} + \frac{100-\bar{y}_н}{M_{возд}}}, \frac{кмоль\ CO_2}{кмоль\ смеси} \quad (2.2)$$

where

$\bar{y}_н$ - is the initial concentration of CO_2 in the mixture;

$M_{возд}$ is the molar mass of air, $\frac{кг}{кмоль}$;

$$y_н = \frac{\frac{6}{44}}{\frac{6}{44}+\frac{100-6}{29}} = 0{,}04\ \frac{кмоль\ CO_2}{кмоль\ смеси}$$

The CO recovery factor$_2$ is calculated using the expression:

$$\alpha = \frac{\overline{y_н} - \overline{y_к}}{\overline{y_н}} \quad (2.3)$$

where

$\bar{y}_к$ - is the final concentration of CO_2 in the mixture.

$$\alpha = \frac{6-0{,}4}{6} = 0{,}933$$

Then the amount of absorbed CO_2 according to formula (2.1) is equal to

$$G = _{CO2}\frac{1{,}05 \cdot 0{,}04 \cdot 44 \cdot 0{,}933}{22{,}4} = 0{,}077\ kg/s$$

Determination of absorbent flow rate

Absorbent flow rate is determined by the formula:

$$L = \frac{G_{CO2} \cdot \varepsilon}{\overline{X_к} - \overline{X_н}},\ \frac{кг}{с} \quad (2.4)$$

where

$\bar{X}_н$ и $\bar{X}_к$ - the initial and final concentrations of CO_2 in the absorbent, expressed in relative mass units, $\frac{кг\ CO_2}{кг\ H_2O}$.

The final concentration of CO_2 in the absorbent is determined by the formula:

$$\overline{X_к^*} = \frac{M_{CO2} \cdot x_к^*}{M_{H2O} \cdot (1 - x_к^*)},\ \frac{кг\ CO_2}{кг\ H_2O} \quad (2.5)$$

where

$x_к^*$ - mole fraction of CO_2 in the absorbent, equilibrium with the initial concentration of CO_2 in the gas phase $\frac{кмоль\ CO_2}{кмоль\ (CO_2 + H_2O)}$, is determined by the expression:

$$x_к^* = \frac{y_н \cdot П}{К},\ \frac{кмоль\ CO_2}{кмоль\ (CO_2 + H_2O)} \quad (2.6)$$

where

P - absorption pressure, kPa;

K - Henry's coefficient for aqueous solutions (tabulated value). For aqueous solution of CO_2 at t = 35° C, Henry's coefficient K = $1.59 \cdot 10^6$ mmHg = 211,947,000 Pa = 211,947 kPa.

We first determine the equilibrium mole fraction of CO_2 in the absorbent:

$$x_к^* = \frac{0{,}04 \cdot 141{,}5}{211\,947} = 26.7 \cdot 10^{-6} \frac{кмоль\ CO_2}{кмоль\ (CO_2 + H_2O)}$$

We then determine the final equilibrium concentration of CO_2 in the absorbent:

$$\overline{X_К^*} = \frac{44 \cdot 26{,}7 \cdot 10^{-6}}{18\,(1-26{,}7 \cdot 10^{-6})} = 65.3 \cdot 10^{-6} \frac{кг\ CO_2}{кг\ H_2O}$$

And determine the absorbent flow rate

$$L = \frac{0{,}077 \cdot 1{,}2}{65{,}3 \cdot 10^{-6} - 0} = 1415\ \frac{кг}{с}$$

Actual CO_2 concentration in the absorbent at the outlet of the absorber:

$$\overline{X_К} = \frac{G_{CO2}}{L},\ \frac{кг\ CO_2}{кг\ H_2O} \quad (2.7)$$

Then the actual concentration of CO_2 in the absorbent at the outlet of the absorber is:

$$\overline{X_К} = \frac{0{,}077}{1415} = 54.4 \cdot 10^{-6} \frac{кг\ CO_2}{кг\ H_2O}$$

Determination of the absorber diameter

We choose the material and type of nozzle [Pavlov, K. F. Examples and problems in the course of processes and apparatuses of chemical technology: textbook for universities. / K.F. Pavlov, P.G. Romankov, A.A. Noskov; ed. by Chl.-Corr. of the Russian Academy of Sciences P.G. Romankov. - 13th ed., 2006]. Raschig rings are simple rings without additional devices. These rings are the cheapest and easiest to produce, they have proven themselves in practice and are the most common type of nozzles.

Table 2.5 - Initial data for calculating the absorber diameter

σ, m /m^{23}	$V_{СВ}$, m /m^{33}	$G_{Нас}$ kg/m^3
204	0,74	532

where

σ - specific surface, m /m^{23} ;
$V_{СВ}$ - free volume, m /m^{33} ;
$G_{Нас}$ - nozzle mass, kg/m^3 .

For further calculations it is necessary to find the equivalent diameter using the formula:

$$d_э = \frac{4 \cdot V_{СВ}}{\sigma},\ м \quad (2.8)$$

$$d_э = \frac{4 \cdot 0{,}74}{204} = 0{,}0145\ м$$

The absorber is calculated for operation at optimum hydrodynamic conditions. The operating gas velocity in the absorber is determined by the formula:

$$\omega = \frac{0{,}6}{0{,}85} \cdot \omega_0,\ m/s \quad (2.9)$$

The gas velocity corresponding to the emergence of the emulsification mode $\omega_0$is determined by the equation:

$$\lg\left[\frac{\omega_0^2 \cdot \sigma \cdot \rho_y \cdot \mu_x^{0,16}}{g \cdot V_{\text{св}}^3 \cdot (\rho_x - \rho_y)}\right] = 0{,}022 - 1{,}75 \cdot \left(\frac{L}{G_y}\right)^{0,25} \cdot \left(\frac{\rho_y}{\rho_x - \rho_y}\right)^{0,125} \quad (2.10)$$

where

ρ_y - average density of the gas phase at operating conditions, $\frac{\text{кг}}{\text{м}^3}$;

μ_x - viscosity of water at operating conditions, mPa · s;

ρ_x - density of water at operating conditions, $\frac{\text{кг}}{\text{м}^3}$;

L - absorbent flow rate, $\frac{\text{кг}}{\text{с}}$;

G_y - gas flow rate, $\frac{\text{кг}}{\text{с}}$.

The average density of the gas mixture is determined by the formula:

$$\rho_y = \rho_{\text{см}}^{\text{ср}} \cdot \frac{T_0 \cdot \Pi}{T \cdot \Pi_0}, \frac{\text{кг}}{\text{м}^3} \quad (2.11)$$

where

$\rho_{\text{см}}^{\text{ср}}$ - average density of the gas mixture at operating conditions, $\frac{\text{кг}}{\text{м}^3}$;

T_0 - temperature, K (273 K is accepted);

T is the temperature required for the absorption process, K;

P_0 = 760 mmHg = 101 300 Pa - atmospheric pressure. For convenience of calculations 101 kPa is taken.

P - pressure required for the absorption process, kPa.

Before determining the average density of the gas phase at operating conditions, the final concentration of CO_2 in the gas mixture, expressed in mole fractions, should be calculated using the formula:

$$y_{\text{к}} = \frac{\frac{\bar{y}_{\text{к}}}{M_{CO2}}}{\frac{\bar{y}_{\text{к}}}{M_{CO2}} + \frac{100 - \bar{y}_{\text{к}}}{M_{\text{возд}}}}, \frac{\text{кмоль } CO_2}{\text{кмоль смеси}} \quad (2.12)$$

where

$\bar{y}_{\text{к}}$ - is the final concentration of CO_2 in the mixture.

$$y_{\text{к}} = \frac{\frac{0{,}4}{44}}{\frac{0{,}4}{44} + \frac{100 - 0{,}4}{29}} = 0{,}0026 \frac{\text{кмоль } CO_2}{\text{кмоль смеси}}$$

CO_2 concentrations at the bottom and top of the column will be determined as: $y^{н}_{CO2} = 0{,}04\,\frac{кмоль\ CO_2}{кмоль\ смеси}$; $y^{в}_{CO2} = y_{к} = 0{,}0026\,\frac{кмоль\ CO_2}{кмоль\ смеси}$.

The average CO concentration$_2$ is determined by the formula:

$$y^{ср}_{CO2} = \frac{y^{н}_{CO2} + y^{в}_{CO2}}{2},\ \frac{кмоль\ CO_2}{кмоль\ смеси} \quad (2.13)$$

$$y^{ср}_{CO2} = \frac{0{,}04 + 0{,}0026}{2} = 0{,}021\,\frac{кмоль\ CO_2}{кмоль\ смеси}$$

The average air concentration is determined by the formula:

$$y^{ср}_{возд} = 1 - y^{ср}_{CO2},\ \frac{кмоль\ CO_2}{кмоль\ смеси} \quad (2.14)$$

$$y^{ср}_{возд} = 1 - 0{,}021 = 0{,}979\,\frac{кмоль\ CO_2}{кмоль\ смеси}$$

The average density of the gas mixture is determined by the formula:

$$\rho^{ср}_{см} = \frac{M_{CO2}}{22{,}4} \cdot y^{ср}_{CO2} + \frac{M_{возд}}{22{,}4} \cdot y^{ср}_{возд},\ \frac{кг}{м^3} \quad (2.15)$$

$$\rho^{ср}_{см} = \frac{44}{22{,}4} \cdot 0{,}021 + \frac{29}{22{,}4} \cdot 0{,}979 = 1{,}31\,\frac{кг}{м^3}$$

We substitute the obtained values and obtain:

$$\rho_y = 1{,}31 \cdot \frac{273 \cdot 141{,}5}{308 \cdot 101} = 1{,}63\,\frac{кг}{м^3}$$

The average molecular weight of the gas mixture is determined by the formula:

$$M_{см} = M_{CO2} \cdot y^{ср}_{CO2} + M_{возд} \cdot y^{ср}_{возд},\ \frac{кг}{кмоль} \quad (2.16)$$

$$M_{см} = 44 \cdot 0{,}021 + 29 \cdot 0{,}979 = 29{,}32\,\frac{кг}{кмоль}$$

The gas flow rate is determined by the formula:

$$G_y = \frac{V \cdot M_{см}}{22{,}4},\ \frac{кг}{с}$$

$$G_y = \frac{1{,}05 \cdot 29{,}32}{22{,}4} = 1{,}37\,\frac{кг}{с}$$

The viscosity of the gas mixture at operating conditions is determined by the formula:

$$\mu =_{y} \frac{M_{см}}{\frac{M_{CO2}}{\mu_{CO2}} \cdot y^{ср}_{CO2} + \frac{M_{возд}}{\mu_{возд}} \cdot y^{ср}_{возд}},\ Pa \cdot s \quad (2.17)$$

where

μ_{CO2} and $\mu_{возд}$ are dynamic viscosity coefficients of CO_2 and air, respectively, at absorption temperature t = 35° C (tabulated value).

$$\mu =_{y} \frac{29{,}32}{\frac{44}{152{,}5 \cdot 10^{-7}} \cdot 0{,}021 + \frac{29}{0{,}019 \cdot 10^{-3}} \cdot 0{,}979} = 1.89 \cdot 10^{-5}\ Pa \cdot s$$

After substituting the corresponding values we obtain:

$$\lg\left[\frac{\omega_0^2 \cdot 204 \cdot 1{,}63 \cdot 0{,}8^{16}}{9{,}81 \cdot 0{,}74^3 \cdot (993{,}6 - 1{,}63)}\right] = 0{,}022 - 1{,}75 \cdot \left(\frac{1415}{1{,}37}\right)^{0{,}25} \cdot \left(\frac{1{,}63}{993{,}6 - 1{,}63}\right)^{0{,}125},$$

whence $\omega_0 = 0.125$ m/s.

Then the operating velocity of the gas will be equal to:

$$\omega = \frac{0{,}6}{0{,}85} \cdot 0{,}125 = 0.09\ m/s$$

The mixture flow rate under operating conditions is determined by the formula:

$$V' = V \cdot \frac{П_0 \cdot T}{П \cdot T_0}, \frac{м^3}{с} \quad (2.18)$$

$$V' = 1{,}05 \cdot \frac{101 \cdot 308}{141{,}5 \cdot 273} = 0{,}85 \frac{м^3}{с}$$

The diameter of the absorber is determined by the formula:

$$D'_К = \sqrt{\frac{4 \cdot V'}{\pi \cdot \omega}}, м \quad (2.19)$$

$$D'_К = \sqrt{\frac{4 \cdot 0{,}85}{3{,}14 \cdot 0{,}09}} = 3{,}47 \text{ м}$$

According to Timonin's handbook A. C. "Fundamentals of design and calculations of chemical-technological and environmental equipment" we take a standard column with diameter $D_К = 3600$ mm.

The actual gas flow velocity is determined by the expression:

$$\omega_Д = \left(\frac{D'_К}{D_К}\right)^2 \cdot \omega, \text{ m/s} \quad (2.20)$$

$$\omega_Д = \left(\frac{3{,}47}{3{,}60}\right)^2 \cdot 0{,}09 = 0.08 \text{ m/s}$$

Determine the irrigation density using the formula:

$$U = \frac{L}{\rho_ж \cdot 0{,}785 \cdot D_К^2}, \frac{м^3}{м^2 \cdot с} \quad (2.21)$$

$$U = \frac{1415}{993{,}6 \cdot 0{,}785 \cdot 3{,}6^2} = 0{,}1399 \frac{м^3}{м^2 \cdot с} = 504 \frac{м^3}{м^2 \cdot ч}$$

The optimum irrigation density is assumed to be $U_{опт} = 5 \frac{м^3}{м^2 \cdot ч}$. At this value, complete wetting of the nozzle is achieved.

Since $U > U_{опт}$, we accept the selected nozzle - ceramic Raschig rings 25 × 25 × 3 mm.

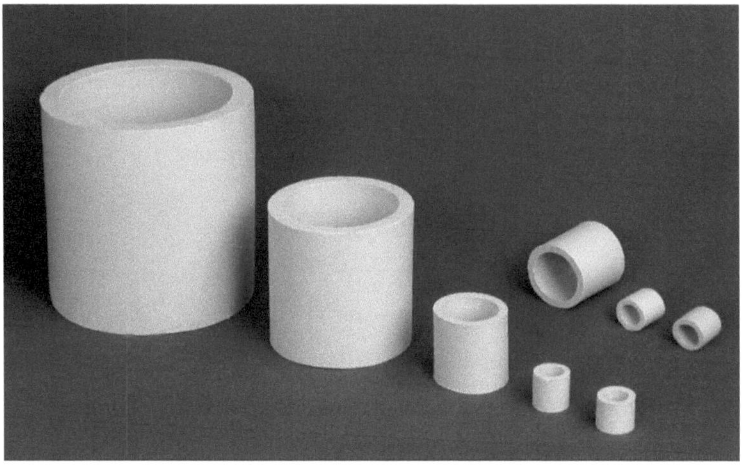

Fig. 2.20 - Ceramic Raschig rings of different diameters

Determination of column height

The height of the nozzle absorber is determined by the formula:

$$H_к = H_н + h_н + h_в + \left(\frac{H_н}{3 \cdot D_к} - 1\right) \cdot 0{,}4, \text{ м} \quad (2.22)$$

where

$H_н$ - nozzle height, m;

$h_н$ and $h_в$ - height of the bottom and top parts of the column, respectively. It is taken depending on the column diameter.

Firstly, it is necessary to determine: the average driving force of the absorption process, the mass transfer coefficient and the nozzle height respectively:

$$\Delta P_{ср} = \frac{\Delta P_н - \Delta P_в}{2{,}3 \cdot lg\frac{\Delta P_н}{\Delta P_в}}, \text{ kPa.} \quad (2.23)$$

where

$\Delta P_н$ - process driving force at the bottom of the column, kPa;

$\Delta P_в$ - driving force of the process at the top of the column, kPa.

$$K_{\Delta P} = \frac{1}{\frac{1}{\beta y} + \frac{\varphi}{\beta x}} \quad (2.24)$$

where

φ - distribution coefficient;

β_y - mass transfer coefficient for gas;

β_x - mass transfer coefficient for liquid.

$$H_н = \frac{G_{CO2}}{M_{CO2} \cdot \sigma \cdot S_к \cdot К_{\Delta P} \cdot \Delta P_{ср}}, м \quad (2.25)$$

where

G_{CO2} - amount of absorbed CO_2 , kg/s;

M_{CO2} - CO molar mass$_2$, $\frac{кг}{кмоль}$;

$S_к$ - cross-sectional area of the nozzle, m2.

σ - specific surface of dry nozzle, m /m^{23} .

Determining the driving force at the bottom of the column

$$\Delta P_н = P_н^{CO2} - P_н^{*CO2}, kPa \quad (2.26)$$

where

$P_н^{CO2}$ - is the partial pressure of CO_2 at the absorber inlet, kPa;

$P_н^{*CO2}$ - is the partial pressure of CO_2 in the gas in equilibrium with the liquid flowing out of the absorber, kPa.

The partial pressure of CO_2 at the inlet to the absorber is determined by the formula:

$$P_н^{CO2} = y_н \cdot П \quad (2.27)$$

$$P_н^{CO2} = 0{,}04 \cdot 141{,}5 = 5{,}66 \text{ kPa}$$

The final concentration of CO_2 in the liquid, expressed in mole fractions, is determined by the expression:

$$x_к = \frac{\frac{\overline{X_к}}{M_{CO2}}}{\frac{\overline{X_к}}{M_{CO2}} + \frac{1}{M_{H2O}}}, \frac{кмоль\ CO_2}{кмоль\ смеси} \quad (2.28)$$

$$x_к = \frac{\frac{54{,}4 \cdot 10^{-6}}{44}}{\frac{54{,}4 \cdot 10^{-6}}{44} + \frac{1}{18}} = 2.23 \cdot 10^{-5} \frac{кмоль\ CO_2}{кмоль\ (CO_2 + H_2O)}$$

The partial pressure of CO_2 in the gas in equilibrium with the liquid flowing out of the absorber is determined by the formula:

$$P_н^{*CO2} = К \cdot x_к, \text{ kPa} \quad (2.29)$$

$$P_н^{*CO2} = 211\,947 \cdot 2{,}23 \cdot 10^{-5} = 4.73 \text{ kPa}$$

Then the driving force at the bottom of the column is:

$$\Delta P_н = 5{,}66 - 4{,}73 = 0.93 \text{ kPa}$$

Determining the driving force at the top of the column

The driving force of the process at the top of the column is determined by the formula:

$$\Delta P_{в} = P_{В}^{CO2} - P_{В}^{*CO2}, \text{ kPa} \quad (2.30)$$

where

$P_{В}^{CO2}$ - is the partial pressure of CO_2 in the gas leaving the absorber upstream, kPa;

$P_{В}^{*CO2}$ - partial pressure of CO_2 in the gas in equilibrium with water, kPa.

The partial pressure of CO_2 in the gas leaving the top of the absorber is determined from the expression:

$$P_{В}^{CO2} = y_{К} \cdot \Pi \quad (2.31)$$

$$P_{В}^{CO2} = 0{,}0026 \cdot 141{,}5 = 0{,}37 \text{ kPa}$$

Since pure water is supplied to the absorber spray, the partial pressure of CO_2 in the gas in equilibrium with water is zero ($P_{В}^{*CO2} = 0$); hence the driving force of the absorption process at the top of the column:

$$\Delta P_{в} = 0{,}37 - 0 = 0.37 \text{ kPa}$$

After substituting the corresponding values, we obtain the average driving force of the absorption process:

$$\Delta P_{ср} = \frac{0{,}93 - 0{,}37}{2{,}3 \cdot lg\frac{0{,}93}{0{,}37}} = 0.61 \text{ kPa}$$

5.3.Determination of the mass transfer coefficient

The mass transfer coefficient $K_{\Delta P}$ is determined by formula (2.23). constant φ in this equation is determined by the formula:

$$\varphi = \frac{K \cdot M_{H2O}}{\rho_x} \quad (2.32)$$

where

K - Henry's coefficient for aqueous solutions, kPa;

M_{H2O} - molar mass of water, $\frac{кг}{кмоль}$;

ρ_x - the density of water under operating conditions, $\frac{кг}{м^3}$.

$$\varphi = \frac{211\,947 \cdot 18}{993{,}6} = 3839{,}62$$

The mass transfer coefficient on the gas side is determined by the formula:

$$\beta_y = \frac{Nu' \cdot D_y \cdot \rho_y}{d_э \cdot P_{иг} \cdot M_{см}}, \frac{кмоль}{м^2 \cdot с \cdot кПа} \quad (2.33)$$

where

D_y - diffusion coefficient of CO_2 in air under working conditions, $\frac{м^2}{с}$;

ρ_y - average density of the gas mixture at operating conditions, $\frac{кг}{м^3}$;

$d_э$ - equivalent nozzle diameter, m;

$P_{иг}$ - average partial pressure of inert gas (in this case air) in the gas mixture, kPa;

$M_{см}$ - the average molecular weight of the gas mixture, $\frac{кг}{кмоль}$.

The Reynolds criterion is determined by the formula:

$$Re_y = \frac{4 \cdot \omega \cdot \rho_y}{\sigma \cdot \mu_y} \quad (2.34)$$

$$Re_y = \frac{4 \cdot 0{,}09 \cdot 1{,}63}{204 \cdot 1{,}89 \cdot 10^{-5}} = 152{,}2$$

The diffusion coefficient at operating conditions is determined by the formula:

$$D_y = D_0 \cdot \frac{П_0}{П} \cdot \left(\frac{T}{T_0}\right)^{\frac{3}{2}}, \frac{м^2}{с} \quad (2.35)$$

where

D_0 is the diffusion coefficient of CO_2 gas in air, $\frac{м^2}{с}$. For CO_2 is taken to be $13.8 \cdot 10^{-6}\ \frac{м^2}{с}$.

$$D_y = 13.8 \cdot 10^{-6} \cdot \frac{101}{141{,}5} \cdot \left(\frac{308}{273}\right)^{\frac{3}{2}} = 11{,}8 \cdot 10^{-6}\ \frac{м^2}{с}$$

The Prandtl criterion is defined by the expression:

$$Pr' = \frac{\mu_y}{\rho_y \cdot D_y} \quad (2.36)$$

$$Pr' = \frac{1{,}89 \cdot 10^{-5}}{1{,}63 \cdot 11{,}8 \cdot 10^{-6}} = 0{,}98$$

The Nusselt criterion is determined by the formula:

$$Nu' = 0{,}027 \cdot Re_y^{0{,}8} \cdot (Pr')^{0{,}33} \quad (2.37)$$

$$Nu' = 0{,}027 \cdot 152{,}2^{0{,}8} \cdot 0{,}98^{0{,}33} = 1{,}49$$

The average partial pressure of the inert gas is determined by the formula:

$$P_{иг} = P \cdot (1 - y_{CO2}^{ср}), \text{ kPa} \quad (2.38)$$

$$P_{иг} = 141.5 \cdot (1 - 0.021) = 138.5 \text{ kPa}$$

Determine the mass transfer coefficient on the gas side:

$$\beta_y = \frac{1{,}49 \cdot 11{,}8 \cdot 10^{-6} \cdot 1{,}63}{0{,}0145 \cdot 138{,}5 \cdot 29{,}32} = 0{,}487 \cdot 10\text{-}6 \ \frac{кмоль}{м^2 \cdot с \cdot кПа}$$

The mass transfer coefficient on the liquid side is determined by the formula:

$$\beta_x = \frac{Nu_x \cdot D_x}{d_э}, \frac{кмоль}{м^2 \cdot с \cdot кПа} \quad (2.39)$$

where

Nu_x - Nusselt criterion in the liquid phase;

D_x - diffusion coefficient of CO_2 in water, $\frac{м^2}{с}$.

Diffusion coefficient of CO_2 in water $D_x = 0.68 \cdot 10^{-9} \frac{м^2}{с}$.

The Nusselt criterion in the liquid phase is determined by the formula:

$$Nu_x = 0{,}00595 \cdot Re_x^{0{,}67} \cdot Pr_x^{0{,}33} \cdot Ga^{0{,}33} \quad (2.40)$$

where

Re_x - Reynolds criterion in the liquid phase;

Pr_x - Prandtl criterion in the liquid phase;

Ga is the Galileo criterion in the liquid phase.

The Reynolds criterion is determined by the formula:

$$Re_x = \frac{4 \cdot L}{\sigma \cdot \mu_x \cdot S_к} \quad (2.41)$$

$S_к$ - cross-sectional area of the column, m^2, calculated by the formula:

$$S_к = 0.785 \cdot D_к^2, \text{ m}^2 \quad (2.42)$$

$$S_к = 0.785 \cdot 3.6^2 = 10.17 \text{ m}^2$$

Then the Reynolds criterion is determined by the formula:

$$Re_x = \frac{4 \cdot 1415}{204 \cdot 0{,}8 \cdot 10^{-3} \cdot 10{,}17} = 3410{,}16$$

The Prandtl criterion is determined by the formula:

$$Pr_x = \frac{\mu_x}{\rho_x \cdot D_x} \quad (2.43)$$

$$Pr_x = \frac{0,8 \cdot 10^{-3}}{993,6 \cdot 0,68 \cdot 10^{-9}} = 1184,05$$

Galileo's criterion is defined by the formula:

$$\mathrm{Ga} = \frac{d_{\text{н}}^3 \cdot \rho_x^2 \cdot g}{\mu_x^2} \quad (2.44)$$

where

$d_{\text{н}}$ - outer diameter of the nozzle element, m. It is assumed to be 0.025 m.

$$\mathrm{Ga} = \frac{0,025^3 \cdot 993,6^2 \cdot 9,81}{(0,8 \cdot 10^{-3})^2} = 236.45 \cdot 10^{6}$$

Finding the Nusselt criterion:

$$Nu_x = 0,00595 \cdot 3410,16^{0,67} \cdot 1184,05^{0,33} \cdot (236,45 \cdot 10^6)^{0,33} = =8137,84$$

Substitute into formula (2.41) and find the mass transfer coefficient on the liquid side:

$$\beta_x = \frac{8137,84 \cdot 0,68 \cdot 10^{-9}}{0,0145} = 382 \cdot 10^{-6} \frac{\text{кмоль}}{\text{м}^2 \cdot \text{с} \cdot \text{кПа}}$$

Substitute all found values and find the mass transfer coefficient:

$$K_{\Delta P} = \frac{1}{\frac{1}{0,487 \cdot 10^{-6}} + \frac{3839,62}{382 \cdot 10^{-6}}} = 0,083 \cdot 10\text{-}6 \frac{\text{кмоль}}{\text{м}^2 \cdot \text{с} \cdot \text{кПа}}$$

Substitute all found values and find the nozzle height

$$H_н = \frac{0,077}{44 \cdot 204 \cdot 10,17 \cdot 0,083 \cdot 10^{-6} \cdot 0,61} = 16,66 \text{ м}$$

The height of the column as a whole is determined by the formula:

$$H_к = 16,66 + 2,5 + 1,4 + \left(\frac{16,66}{3 \cdot 3,6} - 1\right) \cdot 0,4 = 20,77 \text{ м}$$

Hydraulic resistance of the column with nozzle

The hydraulic resistance of the packed column is determined by the formula:

$$\Delta P = \Delta P_{\text{ор}} + \Delta P_{\text{мс}}, \text{ Pa} \quad (2.45)$$

where

$\Delta P_{\text{ор}}$ - resistance of the irrigated nozzle, Pa;

$\Delta P_{\text{мс}}$ - pressure loss for overcoming local resistances in the column, Pa.

The resistance of the irrigated nozzle is determined by the formula:

$$\Delta P_{\text{ор}} = H_{\text{н}} \cdot \left(\frac{\Delta p}{l}\right)_{\text{сн}} \cdot \left(1 + A \cdot \left(\left(\frac{L}{G_y}\right)^{1,8} \left(\frac{\rho_y}{\rho_x}\right) \cdot \left(\frac{\mu_x}{\mu_y}\right)^{0,2}\right)^{0,225}\right) \quad (2.46)$$

where

A - coefficient depending on ω_y u ω_u - working and inversion rates, respectively;

$\left(\frac{\Delta p}{l}\right)_{CH}$ - resistance of dry nozzle per one metre of height, determined by the formula:

$$\left(\frac{\Delta p}{l}\right)_{CH} = \lambda \cdot \frac{\sigma}{4 \cdot V_{CB}^3} \cdot \frac{\omega^2 \cdot \rho_y}{2}, \text{ Pa} \cdot \text{m} \quad (2.47)$$

Since λ = f (Re), when Re > 40, the coefficient of λ will be determined by the formula:

$$\lambda = \frac{16}{Re_y^{0,2}} \quad (2.48)$$

$$\lambda = \frac{16}{152{,}2^{0,2}} = 5{,}86$$

The resistance of the dry nozzle per one metre of height will be:

$$\left(\frac{\Delta p}{l}\right)_{CH} = 5{,}86 \cdot \frac{204}{4 \cdot 0{,}74^3} \cdot \frac{0{,}09^2 \cdot 1{,}63}{2} = 4.87 \text{ Pa} \cdot \text{m}$$

Resistance of the irrigated nozzle for the mobility point:

$$\Delta P_{op} = 16{,}66 \cdot 4{,}87 \cdot \left(1 + 5{,}1 \cdot \left(\left(\frac{1415}{1{,}37}\right)^{1,8} \left(\frac{1{,}63}{993{,}6}\right) \cdot \left(\frac{0{,}8 \cdot 10^{-3}}{1{,}89 \cdot 10^{-5}}\right)^{0,2}\right)^{0,225}\right) =$$

$$=2004.34 \text{ Pa}$$

Pressure loss for overcoming local resistances in the column is determined by the formula:

$$\Delta P_{MC} = \Delta P_{1,6} + \Delta P_{2,3,4,5}, \text{ Pa} \quad (2.49)$$

where

$\Delta P_{1,6}$ - pressure loss to overcome local resistance of gas flow inlet and outlet in the column, Pa;

$\Delta P_{2,3,4,5}$ - pressure loss to overcome local resistances (two inlets and two outlets from the nozzle), Pa.

Pressure loss for overcoming local resistances of inlet and outlet of gas flow in the column is determined by the formula:

$$\Delta P_{1,6} = (\xi_1 + \xi_6) \cdot \frac{\omega_\Gamma^2 \cdot \rho_y}{2}, \text{ Pa} \quad (2.50)$$

where

ξ_1 - sudden expansion coefficient;

ξ_6 - sudden contraction factor;

$\omega_г$ - gas flow velocity in the inlet and outlet pipelines, m/s.

According to the collection "Examples and problems on the course of processes and apparatuses of chemical technology: textbook for universities" edited by K.F. Pavlov, P.G. Romankov, A.A. Noskov (13th ed., 2006) the speed of gas flow is taken within the range from 5 to 20. For convenience we take $\omega_г = 10$ m/s, then the diameter of gas pipelines is determined by the formula:

$$D_г = \sqrt{\frac{G_y}{0{,}785 \cdot \rho_y \cdot \omega_г}}, \text{ м} \quad (2.51)$$

$$D_г = \sqrt{\frac{1{,}37}{0{,}785 \cdot 1{,}63 \cdot 10}} = 0{,}327 \text{ м}$$

According to GOST 8732-78 we choose a pipeline with standard diameter $D_г$ = 400 mm.

Determine $Re_г$ - Reynolds criterion in gas pipelines by the formula:

$$Re_г = \frac{\omega_г \cdot D_г \cdot \rho_y}{\mu_y} \quad (2.52)$$

$$Re_г = \frac{10 \cdot 0{,}4 \cdot 1{,}63}{1{,}89 \cdot 10^{-5}} = 344973{,}56$$

The ratio of the cross-sectional areas of the gas pipeline and the column $\frac{F_г}{F_к}$ is determined by the expression:

$$\frac{F_г}{F_к} = \frac{0{,}785 \cdot D_г^2}{0{,}785 \cdot D_к^2} \quad (2.53)$$

$$\frac{F_г}{F_к} = \frac{0{,}785 \cdot 0{,}4^2}{0{,}785 \cdot 3{,}6^2} = 0{,}012$$

Local resistance coefficients ξ are found according to the collection "Examples and problems on the course of processes and apparatuses of chemical technology: textbook for universities" edited by K.F. Pavlov, P.G. Romankov, A.A. Noskov (13th edition, 2006) - Table XIII. The coefficient of sudden expansion ξ_1 = 0.81 and the coefficient of sudden contraction ξ_6 = 0,45 are taken depending on the ratio $\frac{F_г}{F_к}$ and Reynolds number $Re_г$. Then the pressure loss for overcoming local resistances of the gas flow inlet and outlet in the column will be:

$$\Delta P_{1,6} = (0{,}81 + 0{,}45) \cdot \frac{10^2 \cdot 1{,}63}{2} = 102.69 \text{ Pa}$$

The pressure loss to overcome local resistances - two inlets and two outlets from the nozzle - is calculated according to the formula:

$$\Delta P_{2,3,4,5} = \left(2 \cdot \xi_{2,4} + 2 \cdot \xi_{3,5}\right) \cdot \frac{w_{н}^2 \cdot \rho_y}{2}, \text{ Pa} \quad (2.54)$$

where

$\xi_{2,4}$ - coefficient of sudden constriction at nozzle entry;

$\xi_{3,5}$ - coefficient of sudden expansion at nozzle exit;

$\omega_н$ - is the actual gas flow velocity in the nozzle, m/s.

The actual gas flow velocity in the nozzle is determined by the formula:

$$\omega_н = \frac{\omega}{V_{св}}, \text{ m/s} \quad (2.55)$$

$$\omega_н = \frac{0{,}09}{0{,}74} = 0.12 \text{ metres per second}$$

We determine the Reynolds criterion by the formula:

$$\mathrm{Re}_н = \frac{\omega_н \cdot d_э \cdot \rho_y}{\mu_y} \quad (2.56)$$

$$\mathrm{Re}_н = \frac{0{,}12 \cdot 0{,}0145 \cdot 1{,}63}{1{,}89 \cdot 10^{-5}} = 150$$

Determine the ratio of cross-sectional areas of the gas pipeline and the column $\frac{F_{н.св.}}{F_к}$ by the expression:

$$\frac{F_{н.св.}}{F_к} = \frac{V_{св} \cdot S_к}{S_к} = V_{св} \quad (2.57)$$

$$\frac{F_{н.св.}}{F_к} = 0.74 \text{ m}/\text{m}^{33}$$

Then at the ratio $\frac{F_{н.св.}}{F_к} = 0.74 \text{ m}/\text{m}^{33}$ sudden contraction coefficient at the nozzle entrance $\xi_{2,4} = 0.8$; coefficient of sudden expansion at nozzle exit $\xi_{3,5} = 0{,}8$.

The pressure loss for overcoming local resistances (two inlets and two outlets from the nozzle) is calculated according to the formula:

$$\Delta P_{2,3,4,5} = (2 \cdot 0{,}8 + 2 \cdot 0{,}8) \cdot \frac{0{,}12^2 \cdot 1{,}63}{2} = 0.04 \text{ Pa}.$$

Determine pressure losses for overcoming local resistances in the column:

$$\Delta P_{MC} = 102.69 + 0.04 = 102.73 \text{ Pa} \quad (2.58)$$

Then the total hydraulic resistance of the packed column is equal to:

$$\Delta P = 2004{,}34 + 102{,}73 = 2107{,}07 \text{ Pa}$$

Calculation of auxiliary equipment (gas blower)

The total pressure developed by the gas blower (P, Pa) is determined by the formula:

$$P = 1.05 \cdot (P - P_{am} + \Delta P) \quad (2.59)$$

where

P - absorption pressure, Pa;

P_{am} - atmospheric pressure, Pa;

ΔP - hydraulic resistance of the nozzle column, Pa.

$$P = 1.05 \cdot (141{,}500 - 101{,}325 + 2107.07) = 44396.17 \text{ Pa}$$

The power consumed by the gas blower is determined by the formula:

$$N_{B} = \frac{V \cdot P}{1000 \cdot \eta}, \text{ kW} \quad (2.60)$$

where

η - total efficiency of the gas blower (assumed to be 0.6);

$$N_{B} = \frac{1{,}05 \cdot 44396{,}17}{1000 \cdot 0{,}6} = 77.7 \text{ kW}$$

According to the collection "Basic processes and apparatuses of chemical technology: design manual" under the editorship of Y. I. Dytnersky we select a gas blower of mark TV-50-1,6 with the following characteristics: productivity 1 m^3 /s; pressure - 60 000 Pa; electric motor type AO2-92-2; power - 100 kW. Thus, in the course of calculations the amount of absorbed gas G_{CO2} = 0,077 kg/s was found.

Bibliography

1. Federal Service for Labour and Employment : official site. - Moscow. - URL: https://git87.rostrud. gov.ru/poleznaya_informatsiya/118679.html (date of reference: 03.10.2022). - Text: electronic.
2. Akhtyamov R.G., Meshcheriakova N.A. Analysis of the contribution of the railway industry in the development of global warming / In the collection: PROBLEMS OF SECURITY (SECURITY-2022). materials of the IV International Scientific and Practical Conference dedicated to the 90th anniversary of UGATU. Ufa, 2022. C. 255-260.
3. Akhtyamov R.G., Meshcheriakova N.A. Analysis of negative greenhouse gas emission technologies as ways to implement the strategy of decarbonisation of railway transport / In the collection: PROBLEMS OF SECURITY ASSURANCE (SECURITY-2022). materials of the IV International Scientific and Practical Conference dedicated to the 90th anniversary of UGATU. Ufa, 2022. C. 302-306.
4. Akhtyamov R.G., Meshcheriakova N.A. Ways to implement the strategy of decarbonisation as a way to combat global warming / In the collection: PROBLEMS OF SECURITY ASSURANCE (SECURITY-2022). materials of the IV International Scientific and Practical Conference dedicated to the 90th anniversary of UGATU. Ufa, 2022. C. 341-347.
5. Akhtyamov, R.G.; Senyushkin, N.S.; Sukhanov, A.V. Development of a methodology for identifying potentially dangerous objects / Bulletin of Voronezh State Technical University. 2011. T. 7. № 5. C. 192-197.
6. Elizariev A.N., Akhtyamov R.G., Aksenov S.G., Tarakanov D.A., Tarakanov D.A. Modern technologies for the protection of transport infrastructure objects based on the modelling of dangerous situations / Life Safety. 2018. № 10 (214). C. 23-28.
7. Akhtyamov R.G. Modelling of the stressed state of the technological equipment / In Collection: Technosphere and ecological safety on transport (TEBTRANS-2014). Materials of the IV International Scientific and Practical Conference. 2014. C. 16-19.
8. Titova T., Akhtyamov R., Nasyrova E., Elizaryev A. Methodical approaches for durability assessment of engineering structures in cold regions / In

Proceedings of TRANSOILCOLD 2019. Proceedings of TRANSOILCOLD 2019. (Series: Lecture Notes in Civil Engineering, volume 49). Singapore, 2020. C. 473-478.

9. Akhtyamov R.G. Problems and prospects of technosphere safety on railway transport / Reports of Bashkir University. 2017. T. 2. № 3. C. 433-437.

10. Titova T.S., Akhtyamov R.G., Nasyrova E.S. Technical solutions to prevent thermal pollution of water bodies within the urban landscape / Bulletin of scientific research results. 2016. № 3-4 (20-21). C. 6068.

11. Climate change: causes and consequences. - Text: electronic //. EcoEnergy. OpenEnergy. - URL: https ://www. ekoenergy.org/en/extras/climate-change/ (date of reference: 23.10.2022)

12. The negotiation process and meetings. Paris Agreement. - Text: electronic // United Nations Climate Change. News portal. - URL: https://unfccc.int/ru/peregovornyy-process-i-vstrechi/parizhskoe-soglashenie/chtohttps://unfccc.int/ru/peregovornyy-process-i-vstrechi/parizhskoe-soglashenie/chto- takoe-parizhskoe-takoe-parizhskoe-soglashenie (date of reference: 05.10.2022)

13. Global warming is linked to human activity. - Text: electronic // United Nations : official website. - URL: httpshttps://news.un.org/ru/story/2021/08/1407862news.un.org/ru/story/2021/08/1407862 (date of access: 31.10.2022)

14. The six most frightening consequences of global warming. - Text: electronic // RBC. News portal - URL: httpshttps://trends.rbc.ru/trends/green/61c4bcf99a79471709fb2143trends.rbc.ru/trends/green/61c4bcf99a79471709fb2143 (date of reference: 03.10.2022)

15. The potential of forests in achieving carbon security. - Text: electronic // Russian Academy of Sciences : official website. - http://www.ras.ru/news/shownews.aspx?id=eeb64fd9-ee04-4fa9-8cfehttp://www.ras.ru/news/shownews.aspx?id=eeb64fd9-ee04-4fa9-8cfe-

abd2a26626d&print1abd2a26626d&print=1 (date of reference: 17.10.22)
16. Global movement to save forests. - Text: electronic // United Nations : official website. - URL: https://www. unep.org/en/novosti-i-iistorii/istoriya/v-ramkakh- globalnogo- dvizheniya-za-spasenie-lesov (date of address: 17.10.22)
17. Fighting climate change with determination. - Text: electronic // Food and Agriculture Organisation of the United Nations : official website. - URL: httpshttps://www.fao.org/fao-www.fao.org/fao-stories/article/en/c/1393875/ (accessed 17.10.22)
18. Belstu : electronic library system : website. - Minsk, URL: https://elib. belstu. by/bitstream/123456789/9738/1/lk-18. pdf (date of reference: 17.10.22)
19. How many trees you need to plant to make a difference. - Text: electronic // NakedScience. Infomation Portal. -URL: https://nakedhttps://naked-science.ru/article/biology/v-chem-neprav-trilon-mask-derevya-ne-spasut-planeturu/article/biology/v-chem-neprav-trilon-mask-derevya-ne-spasut-planetu- ot-globalnogo-ot-globalnogo-potepleniya (date of address: 17.10.22)
20. Bury or absorb: Iceland has learnt to permanently remove carbon dioxide from the atmosphere. - Text: electronic // Ecology of Russia. News portal. - URL: https://ecologyofrussia.ru/otkachka-uglekislogo-gaza-izhttps://ecologyofrussia.ru/otkachka-uglekislogo-gaza-iz- atmosferyatmosfery/ (date of address: 18.10.22)
21. Negative emission technologies: What role in meeting Paris Agreement targets? - Text: electronic // EASAC : official website. -URL: https: //easac. eu/fileadmin/PDF_s/reports_statements/Negative_Carbon/EASAC_Report_on_Negative_Emission_Technologies.pdf (date of access: 18.10.22).
22. What is carbon neutrality. - Text: electronic // RBC. News portal . - URL: httpshttps://trends.rbc.ru/trends/green/5ffd5a099a7947594de716cetrends.rbc.ru/trends/green/5ffd5a099a7947594de716ce (date of reference: 21.10.22)
23. Wiley Online Library : digital library system : website. - URL: https://onlinelibrary.wiley.com/doi/full/10.1111/gcbbhttps://onlinelibrary.wiley.

com/doi/full/10.1111/gcbb. 12643 (accessed 23.10.22)
24. Powerful industrial coal-fired boilers - design and operation. - Text: electronic // Autonomous heat. Open Society. URL: https://avtonomnoeteplo.ru/otopitelnye__kotly/188-avtomaticheskiehttps://avtonomnoeteplo.ru/otopitelnye__kotly/188-avtomaticheskie- promyshlennye-kotly-na-promyshlennye-kotly-na-ugle.html (date of reference: 24.10.22)

More Books!
yes

MIX
Papier aus verantwortungsvollen Quellen
Paper from responsible sources
FSC® C105338

Printed by Books on Demand GmbH, Norderstedt / Germany

State of breeding work with vegetab

Almas Asadova

State of breeding work with vegetable and grain beans

In Azerbaijan

ScienciaScripts

Imprint
Any brand names and product names mentioned in this book are subject to trademark, brand or patent protection and are trademarks or registered trademarks of their respective holders. The use of brand names, product names, common names, trade names, product descriptions etc. even without a particular marking in this work is in no way to be construed to mean that such names may be regarded as unrestricted in respect of trademark and brand protection legislation and could thus be used by anyone.

Cover image: www.ingimage.com

This book is a translation from the original published under ISBN 978-3-659-85879-6.

Publisher:
Sciencia Scripts
is a trademark of
Dodo Books Indian Ocean Ltd. and OmniScriptum S.R.L publishing group

120 High Road, East Finchley, London, N2 9ED, United Kingdom
Str. Armeneasca 28/1, office 1, Chisinau MD-2012, Republic of Moldova, Europe
Printed at: see last page
ISBN: 978-620-2-05657-1

TABLE OF CONTENTS

INTRODUCTION	**4**
CHAPTER 1	**7**
CHAPTER 2	**28**
CHAPTER 3	**30**
CHAPTER 4	**59**
CHAPTER 5	**67**

Providing food for the world's population is the most pressing issue of the modern period. The problem of nutrition of mankind is of great importance nowadays because one third of the world's population is hungry, especially on the African continent. Therefore, the Governments of almost all states and international organisations are dealing with this problem. Special organisational and economic measures are taken, in particular, such as stimulation of grain production, exchange of breeding material in order to increase the efficiency of the process of creating new varieties, enrichment of genoplasm as a basis for them.

The issue of supplying quality food to a rapidly growing population is acute around the world .

Agricultural institutes in different countries are joining together in centres to accelerate the development of early maturing, high-yielding, disease- and drought-resistant varieties of cereals and legumes. This helps to accelerate breeding and seed production and to disseminate new varieties more widely in countries with similar climates.

One such centre is the International Centre for Agricultural Research in the Dry Areas - ICARDA. Mission - ICARDA contributes to improving the welfare of poor people living in drylands by enhancing food security and reducing poverty through research and co-operation to achieve sustainable growth in agricultural production and income while ensuring efficient and equitable use and conservation of natural resources. ICARDA's activities are aimed at conducting advanced agricultural research; ensuring the free exchange of germplasm and information needed for research.

Over the last 17 years, ICARDA has distributed about 50 varieties of cereals and leguminous crops in eight countries of Central Asia and the South Caucasus. They have been successfully released and introduced into agricultural practices in these countries.

The organisation was founded at a conference in Quebec on 16 October 1945. FAO acts as the lead agency for rural development and agricultural production in the UN system. The motto of the organisation is "helping to build a world without hunger". FAO works to alleviate poverty and hunger in the world by promoting agricultural development, improving nutrition and addressing food security - the availability to all, at all times, of the food needed for an active and healthy life.

Another important component of FAO's work is the conservation of agricultural genetic resources and the biodiversity of plant and animal species used by humans. FAO considers biodiversity as a prerequisite for food and agriculture and considers it the most important resource of the earth.

FAO STAT is a public product provided by FAO to users worldwide.

Recently, close links have been established with the newly independent CIS countries. The IWWIP programme ensures that cooperating countries test different varieties of cereals and legumes in contrasting environments and select the best ones adapted to a very wide range of environments.

At all times, the determining international standard of socio-economic development of a civilised state has been the standard of living of its population, the main indicator of which is the level of consumption of vital foodstuffs; including the consumption of vegetables and fruits. In connection with the above-mentioned, many countries have adopted and are successfully implementing state programmes for the development of vegetable growing.

The main source of vegetable protein for human nutrition has long been grain legumes, among which beans occupy a special place. Vegetable beans proteins are close to meat protein, which beans have more than in fish and meat, and due to the contained mineral salts of calcium, iron, vitamins and other substances necessary for the human body, vegetable beans are recommended for liver diseases, avitaminosis, and others.

The relevance of this work is primarily related to the problem of creating new varieties of grain legumes, combining consistently high productivity with environmental resistance to extreme environmental conditions (abiotic and biotic).

INTRODUCTION

All grain legumes belong to the legume family *(Fabaceae)* and have much in common in terms of plant biology, cultivation methods and quality of products. Legumes can be divided with some approximation into 3 groups according to their botanical characteristics: 1) plants with pinnate leaves and seedpods remaining in the ground - peas, chinas, chickpeas, chickpeas, lentils, beans and vetch; 2) plants with triple leaves and seedpods coming out (greening) - beans, soybeans, dolichos (peanuts); 3) plants with flannel-shaped leaves, with seedpods also coming to the surface - lupins.

Grain legumes include: peas, fodder beans, vetch, common bean, lupin, soya bean, lentils, chickpea, chickpea. They are cultivated to produce seeds with high protein content. These crops are divided by their economic importance into: food, fodder, technical and universal. Leguminous crops are of great national economic importance due to their diverse use and especially because they are the most important sources of vegetable protein. Cultivation of these crops contributes to the increase of grain resources in the country, successful solution of the problem of protein fodder and improvement of soil fertility through fixation of air nitrogen.

Grain legumes play a major role in human nutrition and in the feeding of farm animals. The protein of seeds of these crops contains almost all essential amino acids: tryptophan, lysine, methionine, valine, threonine, phenylalanine, leucine, isoleucine, - necessary for growth and development of a living organism. The following amino acids are also essential: histidine, arginine, cystine and trosine. Lack of essential amino acids in food leads to metabolic disorders in the body and to disease.

Grain legume seeds have 2-3 times more protein than cereal seeds.

A characteristic feature of legume protein is that it is easily soluble in water, in solutions of neutral salts and in weak solutions of alkalis. The more easily the protein dissolves, the more accessible it is to the human and animal organism.

Beans and lentils are characterised by high flavour and culinary qualities, they are used only in human nutrition. In world farming, grain legumes occupy about 13-14% of grain bread sowing. In terms of sown area, peas and soya take the first place, followed by lupine. Beans, lentils, chickpeas and fodder beans are cultivated in small areas.

Beans are cultivated for grain, as vegetables and for canning. Mature seeds, green spathes, immature seeds of vegetable bean varieties are consumed. Vegetable beans are valued for their excellent flavour, nutritional and medicinal properties.

Beans enrich the soil with nitrogen and improve its physical properties. After harvesting beans, 40-60 kg of nitrogen per 1 ha remains in the soil.

The main feature of modern farming is the production of crop production with limited anthropogenic energy consumption and preservation of the environment from degradation and pollution. One of the ways to address these issues is the introduction of new varieties into production, agrocenoses of which, due to their significant adaptive potential, provide a high level of productivity realisation at minimum energy costs and have a positive impact on soil fertility elements [53; 66].

Climate changes caused by natural phenomena and anthropogenic pollution of the environment lead to a decrease in immunity and adaptive properties of existing varieties of cultivated crops, which fundamentally changes the focus of breeding. In modern conditions, an expedient and economically justified direction of breeding is to obtain varieties for specific conditions of a particular region. One of the main tasks of breeding is the development of adaptive-ecological direction, which allows expanding the adaptive capabilities of new varieties in their geographical distribution [38].

Multiplication and rapid introduction of high-yielding varieties into production is an effective means of increasing yields [61]. One of the main reasons preventing the widespread use of beans is their suitability for mechanised harvesting. Such varieties should have high anchorage of lower beans, bean resistance to cracking, determinant type of growth, simultaneous ripening, resistance to lodging [15;47;71].

E.E. Ditmer came to the conclusion through his research that for grain direction beans the bush should be compressed, without a tendency to strong branching [36]. Some breeders believe that single-stemmed forms of beans can be promising [72]. All varieties of grain use have a peculiarity of technological cultivation of beans for grain with mechanised harvesting.

Creation of varieties with a complex of certain selectable traits for specific soil and climatic conditions will ensure the development, efficiency and sustainability of agro-ecosystems.

The method of modelling allows to select traits and forms significant in the breeding process to include them in the creation of new bean varieties when implementing the ecological approach of breeding.

Soybean ranks first in terms of sown area in world farming, followed by beans in second place, groundnut in third, chickpea in fourth, pea in fifth, followed by beans, vigna, etc.

CHAPTER 1

LITERATURE REVIEW

1.1. The importance of source material in breeding and ways of its optimisation in connection with the task of increasing the efficiency of the breeding process

The main task of breeding, as a science, is to create new varieties that exceed the best of the existing varieties in their economic qualities. The success in the development of new varieties depends to a large extent on the correct choice of source material. Involvement of a wide and diverse source material and its comprehensive study are the primary prerequisite for the correct setting of breeding work. The main sources of source material, as is known, for breeding are: 1) the best local varieties - populations; 2) the best breeding varieties; 3) world collections of cultivated plant forms, collected in national repositories of different countries of the world, where samples of wild and semi-cultivated relatives of the most important species of cultivated plants are concentrated.

The doctrine and concept of source material for breeding was first given by N.I. Vavilov. A huge gene pool on beans was collected directly by N.I.Vavilov and his collaborators in most countries of the world. At present, the collection of beans includes 6563 samples, of which vegetable beans more than 2000 samples. The old vegetable bean varieties in the collection are still in demand today, as they often carry valuable traits. Many vegetable varieties of the USA due to their productivity, resistance to common mosaic, bacterioses, anthracnose, suitability for mechanised harvesting have become world famous - Tenderette, Corbet Refugee, Giant Stringless Green Pod, Blue Lake, Resistant Cherokee, Cornell 49-242, Idaho Refugee, Tenderwhite, Logan, Contender and others. The vegetable variety Corbett Refugee served as the basis for a number of high-quality varieties resistant to common mosaic (Harvester, White Seeded Tendercrop, Excutive, etc.). The Blue Lake group varieties are still the benchmark of high bean quality [105].

The high popularity of vegetable beans in Bulgaria has led to the development of Zarya, Oreol, etc. varieties resistant to common mosaic and with high bean quality. In Germany, mainly vegetable varieties are grown, including the long-known Saxa, Wilhelm, Triumph, as well as disease-resistant varieties Declivis Romulus, Declivis Remus, Valja, Lusia, Jutta, Esto, Dilana.

The main part of the collection (97%) is common bean. The VIR bean collection gave a start to the development of breeding work in the countries of the former USSR and in Russia. Comprehensive study of the VIR collection contributes to the constant expansion of the geographical range of the crop as a whole. Over the last decade, the collection has received more than 500 breeding and local varieties from all over the world [10].

The main objectives of bean breeding at present in all growing areas is the creation of high-yielding varieties with stable yields over the years, resistant to diseases, pests and adverse environmental factors, suitable for mechanised cultivation, as well as having high quality beans in vegetable bean varieties. Targeted supply of breeders with source material is facilitated by the established trait collections on the most valuable breeding-important traits. Breeders have achieved considerable success in the development of new varieties and new breeding methods, but many problems remain. At the modern level of theoretical research of bean biodiversity the problems of bean breeding are quite solvable. Even N.I.Vavilov pointed out that the success of breeding work ... is largely determined by the source material [13]. The availability of a variety of source material is necessary for the creation of new varieties or improvement of currently used varieties, for the implementation of various breeding programmes. The richest collection of bean accessions, concentrated in the VNII of Plant Industry named after N.I. Vavilov (VIR), has such source material [10].

In order to develop a new variety, it is necessary to know the conditions of one's zone. Namely soil types, meteorological factors, diseases, agrotechnical peculiarities, level of farming culture, limiting environmental factors. All this is important for the possibility of creating the most adapted to local conditions varieties. For any conditions, the best variety will be the one that has a complex of economically useful properties and the least number of negative traits [59;12].

1.2. Economic importance and utilisation, modern production

Beans (genus *Phaseolus* L.) are cultivated for seed as a protein legume, as a vegetable crop and, less frequently, as a flowering plant and for green fertiliser. Beans are widely known and popular on all continents of the globe. This crop plays an important role in eliminating the deficiency of complete protein in the human diet. Bean seeds contain 17.3-23.4 per cent [9]. It is important that its amino acid composition is close to meat and fish proteins, and its digestibility by humans reaches 75-80% [70]. High nutritional value. Bean seeds are 3.5 times more calorific than potatoes. The unripe bean and shifts harvested in the phase of technical ripeness (on the blade) contains 9-11% dry matter, 2.4-3.4% sugar, 1.6-2.8% protein, 20-30mg% ascorbic acid [80]. Beans also contain vitamins K, Bi, B2, PP, Bb, pantothenic acid, minerals - sodium, potassium, magnesium, calcium, iron, phosphorus, iodine.

Along with exclusively ornamental species, which are highly valued for the beauty and fragrance of flowers (P. *caracalla)*, and species whose nutritional value is reduced to digging up and consuming tuberous-thickened roots only in periods of hunger *(*P. tuberosus*)*, there are species used exclusively for green fodder *(P. retusus),* for green fodder and green fertiliser *(P. semirectus).* The majority of species have complex use: for grain, straw, green fodder and green fertiliser (P.*trilobus, P.aconnitifolius, P.mungo, P.aureus, P.angularis, P.calcaratus, P.acutifolius, P.multiflorus).* For 2 species *(P.vulgaris, P.lunatus*) the main purpose is food (grain, green spathe) and industrial (canned grain and green spathe); a lesser role is played by fodder and sideration.

In world agriculture, common bean occupies a dominant position among food grain legumes and is characterised by the greatest polymorphosis of traits and properties. Particularly great is the diversity of beans in size, shape and colour of seeds. There are differences in plant architectonics: height, type of growth, shape and density of the bush, the level of arrangement of beans on the stems, as well as the duration of the vegetation period. Botanical classification of common bean is based on the ecology of the source material. 30 ecotypes have been established [42].

Beans give a great polymorphism of biological forms due to the range of habitats they occupy. Thus, in tropical forests, they are represented by typical perennial climbing vines, whose height reaches the height of the first-tier trees, with woody stems split like strands of rope, branched only at the top, with powerful foliage located, like flowers, in the upper tiers. In more sparse forests and along edges, the foliage of lianas is dense and uniform; the mass of flowers and beans formed along the entire length of the stems is striking. In shrub associations, vines do not reach such lengths; along with perennials, there are also annual species: their height does not exceed a few metres, their stems are weakly woody or herbaceous. In keeping with the support, the height of which is often lower than the liana, and in the progressive movements to upholster a new support, the stems often have to flail on the ground or flip from bush to bush. The stems therefore branch low and the foliage is mainly located at the base.

In transition to steppe, grassy and more open habitats (screes, rock formations), in parallel with increasing sparseness of the grass stand and decreasing height of its components, which can play the role of supports for climbing stems of fasols, the latter gradually begin to lose the ability to twist. The bush becomes ascending, densely branched from the base, branches - weakly curling or wavy, touting leaves at the top. In savannahs, as well as when living among rocks, there are already perennial semi-shrubby forms with semi-erect woody stems. On loose substrates in semi-deserts and on sandy coasts, in order to protect from winds and drying out from strong insolation, bean forms with thick straight nodes grow. The leaves of these forms are coarse, grey or yellow-green, densely pubescent.

The centres of bean polymorphism in the Americas coincide with the centres of origin of agriculture. According to Spinden, bean culture originated 4000 years B.C., and according to Bukasov - 3000 years, in the area of dry farming with limited use of irrigation on the South Mexican and Guatemalan plateaus. S.M.Bukasov believes that the bean culture was introduced by the Protomaians who lived in the Guatemalan highlands [35].

It is one of the main vegetable crops in many countries of the world. The leading consumers are Asian countries, primarily China, Indonesia and Turkey. Asia is home to 72.7 per cent of the world's vegetable bean acreage. The area planted with this crop in the world is increasing annually, and in 2010, according to FAO data, it totalled 1 million 495 thousand ha (FAO STAT) [105].

Depending on the shape of the bean in cross-section, the presence of a parchment layer in the flaps and fibre in the bean seam, vegetable varieties are divided into sugar, semi-sugar and tussock varieties. The best sugar varieties have fleshy, rounded beans with no parchment layer in the green bean flaps and no fibres in the bean seams. Semi-sugar varieties may have a small amount of fibre in the bean flaps. Turshe beans are characterised by their flat shape and the absence of a parchment layer and fibre in the beans. Beans with long thin beans that are rounded in cross-section are called asparagus beans. It is known from bean genetics that the shape of the bean depends on many genes. Plants with genes Ea, Ebja, lb have flat beans, and with ea, eb, ia, ib- rounded. The recessive gene st determines the presence of fibre in bean flaps. The Ts gene determines the dependence of fibre formation on temperature - fibre develops to a greater extent at high temperature. The presence of a parchment layer in the bean leaflets is controlled by the gene To.

As a row crop, beans are a good precursor to spring cereals and other crops.

Many bean species are common in household crops. In the Caucasus and Central Asian republics, mung beans and adzuki beans are cultivated. Here, beans are often cultivated in mixed crops with maize, potatoes and melons [10].

1.3. Botanical characterisation and biological features

There are 230 bean species in nature, of which 17 are used in culture. In the Soviet Union, five species are of economic and production importance: common, multiflower, tepary, lima and mung bean. The bean varieties for grain in the USSR belong to two species: common and mung bean.

The common bean is a well-known vegetable garden plant. The name of the plant is derived from the Greek "narrow, long boat" - from the shape of the open flaps of the bean fruit.

The common bean (Phaseolus *vulgaris* L.) was cultivated in Europe for a long time as an ornamental plant, and only in the 17th century it became known as a food crop. This plant has been known to mankind since very ancient times. The earliest mention in ancient written records of the bean dates back to the II millennia BC. It was used as food in ancient China. And the first bean seeds were discovered by archaeologists working on excavations of monuments of pre-Inca culture in Peru. This plant was widespread in the ancient Incas and Aztecs, and the ancient Greeks and Romans used beans not only for food, but also as a medicinal remedy. Slavs got acquainted with beans around the XI century. There are American and Asian types of beans. American species are characterised by large flat beans with long beaks and large seeds. Asian species are small cylindrical beans. The common bean is the most widely cultivated species. In terms of growth pattern there are forms from bushy to curly, in terms of growing season length from early maturing to late maturing. It has forms of tall, semi-hanging, bushy and bushy with curling (nutting) with a stem height of 40-250 cm.

The root is tap root, penetrating to a depth of more than 1 metre.

The stem of some varieties is curly, of others - straight; strongly branched, covered with sparse hairs.

Bracts small, ovate; bracts linear-lacetate.

Leaves are triply compound, pinnately pedunculate, on long petioles. True leaves are triple, primary leaves are paired; leaflets are large, ovate, broadly ovate or rhomboidal-ovate in shape; often acuminate; colouring is yellow-green, green, dark green and anthocyanine. Stems and leaves are pubescent.

Flower stalks are axillary, with 2-6 flowers each. Flowers are greenish-white, pink, dark pink and violet, five-petalled in structure: a sail, two wings and two fused boat petals, moth-like, collected in axillary **cysts.** Stamens are ten.

Fruits are beans, hanging, 5-20 cm long, 1-1.5 cm wide, straight or curved, flattened or almost cylindrical, pale yellow and green to dark purple, with two to eight (according to others, three to seven /seeds).

Seeds 5-15 mm long, elliptic, white to dark purple and black, unicoloured or mosaic, mottled, spotted.

Beans are sword-shaped, sabre-shaped, straight, smooth, wrinkled (in sugar varieties), clear-shaped. The bean structure distinguishes between husking and sugar bean varieties. In husking varieties, the bean walls have a parchment layer. Sugar varieties have no parchment layer and semi-sugar varieties have a weak or late-developing parchment layer. The sugar forms are called asparagus beans. The colour of the beans is yellow or green when immature, and predominantly straw yellow when mature. Seeds are ovoid (elliptic), ovate (elliptical), valvate, kidney-shaped, spherical, and other various combinations of these shapes. The colour of seeds is different. The seeds are small (1000 seeds weight up to 150 g), medium (1000 seeds weight from 150 to 450 g) and large (1000 seeds weight 450-750 g).

Common bean and lima bean are facultative self-pollinators. The incidence of natural cross-pollination in common bean increases as it approaches the equator. Authors' opinions on the propensity of common bean to natural hybridisation vary widely. In Sweden, according to Lamprechts, cases of natural cross-pollination are extremely rare (0-0.3%). Schiemann (Schiemann) for Germany accepts 12% of cases of natural cross-pollination.

For Georgia, L.L. Dekaprelevich considers natural hybridisation to be about 5%. Under the conditions of dry year 1935, we had to observe the phenomenon of cleistogamy for a number of common bean ("Hodson") forms in Krasnodar [35].

The flowering of the bean tassel comes from the bottom upwards. In bush forms, the terminal tassel blooms earlier and the lateral tassels, even if they are lower down, somewhat later. In common curly beans of temperate latitudes, flowering starts at nodes IV-VII and proceeds basipetally upwards on the bush. In high-carried beans of the Caucasus and Latin America flowering is acropetal:

it starts in the upper third, less often in half of the bush and from here it spreads in both directions, upwards and downwards. One brush blooms in common beans for 10-14 days, in multiflowered beans for 11 days; a whole bush plant blooms for 20-26 days, while a climbing plant blooms for 1-3 months.

Beans do not tolerate heavy cold, clayey and silt soils with high groundwater table. Stagnation of surface water for 2-5 days causes either direct death of the plant or its strong oppression and, as a consequence, death from fungal diseases. In the subtropics, beans do well on heavy clay soils of slopes, heavy alluvial soils of the coast and red soils. In northern regions, beans grow better on light loamy soils. Loams, chernozems, marl and calcareous soils are the best for bean culture in temperate countries. On dry sandy, stony and peaty soils it greatly reduces yields. The multiflower bean tolerates wetter soils better than other species.

In relation to soil salinity, according to Hendry (1918) for California, P.lunatus L. is the most salt tolerant species, below it is tepary. Common bean is the least salt tolerant, but gives a large amplitude of resistance for individual forms. Pea-shaped ecotype ("Small White") turned out to be the most resistant of the tested varieties.

Beans are heat-loving crops. In the sprouting phase it cannot tolerate negative temperatures even 1-2^0 C. It is sown in late dates. Only some varieties in the phase tolerate short-term spring frosts up to 3^0 C. At 2-3^0 C the bean leaves lose their green colouring and turn yellow. Lack of heat and damp, rainy weather during flowering causes the flowers to fall off. Beans require slightly higher average daily temperatures for their development than peas and lentils. According to V.N.Stepanov (1950), bean seeds germinate at a temperature of 10-12^0 C, and sprouts appear at 12-13°C. Bean sprouts cannot withstand prolonged temperature decrease and die at -1, -1.5°C (at plant level) [86].

In dry soil, bean seeds, if they have not had time to swell, can germinate for a long period of time and sprout after rain or irrigation. If the germinated seeds are in a dry soil layer, they may dry out and fail to germinate.

Different bean species and varieties react differently to the length of the day. All wild bean species are short-day plants, while cultivated species, as numerous studies conducted in our country have shown, react differently to day length depending on the variety. According to A.V.Doroshenko, V.I.Razumov, P.I.Kovaleva, A.S.Vasiliev, A.M.Pavlova and others, can be divided into 4 groups: 1) sharply short-day plants; 2) plants reacting to some extent to the length of the day, but remaining within the range from short-day to neutral; 3) plants weakly reacting to the shortened day, remaining within the range from neutral to long-day; 4) long-day plants [7].

The fourth group unites mainly vegetable varieties of common bean and varieties of multiflowered bean (variety Lopata). Varieties of this group normally flower and fruit under conditions of natural long day.

Beans are one of the most light-demanding plants. Different bean species and varieties react differently to the length of the day. It has a special sensitivity to light at a young age, when at insufficient intensity of light illumination of yield. In the period after the beginning of flowering, the demand for light is reduced, so in southern areas it is advisable to even lightly shade plants. A number of works by Russian researchers (P.I.Kovaleva, A.S.Vasilieva and I.M.Vasiliev) put forward the position that there is no reaction to yarrowing, as its first stage is very short and takes place at a wide amplitude of temperature fluctuations. Experiments of P.I.Kovaleva showed specificity of ecotypes and forms to the conditions of passing the stage of yarrowing: so, northern forms better pass the stage of yarrowing at 8°C, southern forms at 25°C. Beans require darkness and increased temperature to pass the second, light stage.

Beans are short-day plants. However, there are varieties that are neutral and even respond favourably to a long day [48]. In the south, almost exclusively short-day varieties and partly neutral varieties are common. In the north, only neutral and long-day varieties can be cultivated. Garnery and Allard found that the bean is a typical short-day plant in terms of its response to day length. According to P.I. Kovaleva, beans can be divided into 3 groups according to their photoperiodic reaction:

1. Sharp-short day group with flowering only on 10 and 14-hour days. It includes Asian beans (P. ca/caratus, *A. aconitifolius},* flowering on the 14-hour day with a delay of 5 days against the 10-hour day, and of *P. vulgaris*, South and Central Amercano and Mexican forms, flowering on the 14-hour day with a delay of 10-20 days.
2. Group with flowering on 10, 14, 18 and 24 hour day with delay in flowering against 10-hour day by 10-25 days. This group includes forms of *P.vulgaris* from the south and central part of the USSR and partly forms from Mexico, Guatemala, Colombia and Bolivia.
3. A group of forms flowering on 10, 14, 18 and 24-hour days, with a delay in flowering against the 10-hour day by 0-12 days. For example, the variety

"Rosovo-pestrya" from the Steppe Experimental Station (Voronezh Region) blooms in the same terms at any length of the day. This is a rather motley group of *P.vulgaris* forms, including USSR, Central European and Mexican varieties. It also includes a variety of forms of *P.multiflorus,* which gives a delay in flowering on a 24-hour day by 0-2-9 days against a 10-hour day and includes forms of weakly reactive, neutral and long-day forms.

Common bean is a moisture-loving plant. Seed swelling requires 104.5% of water from seed weight, soil moisture should be 60-80% of field water capacity. Beans do not tolerate air drought well. Bean flowers and young ovaries suffer most of all from air drought, especially badly tolerated drought during the period of seed filling. Yields drop sharply in drought years. A strong negative effect of drought on bean yields was observed at the Kuban experimental station of VIR, located in the steppe zone [7]. In dry years, bean yields here averaged 5.6 c/ha, while in years with moisture they reached 17.4 c/ha.

Excessive moistening has a detrimental effect on beans if it is accompanied by a drop in air temperature. Sufficient and excessive moisture in autumn delays seed ripening. Bean seeds ripening in damp cool conditions contain less protein, have more intensive respiration, and lose germination faster compared to seeds obtained in relatively dry autumn conditions [2].

Bush bean varieties finish flowering within 2 to 3 weeks and mature amicably in 30 to 45 days.

The vegetation period of common CIS bean varieties averages 80-100 days or more, but as a late-sown crop, it matures much later than peas.

All varieties of common bean are unstable to soil and air irrigation. Satisfactory yields are obtained only in areas with annual precipitation of at least 450-500 mm in the absence of prolonged droughts and dry winds in the phases of flowering, setting and filling of beans. It tolerates excessive moisture well, but without temperature reduction.

Good bean yields are formed on black earth and non-black earth soils with medium cohesion, loamy, marly, rich in known soils. Heavy loamy, cold, silty soils with high standing of subsoil water, as well as saline soils are not suitable for bean cultivation.

1.4. **Production technology bean crop**

In the production technology of advanced farms that receive consistently high yields of beans, special attention is paid to such agrotechnical measures as optimal sowing time (the most favourable conditions for high field germination are created at a soil temperature of 15-200C and 25-30% moisture of absolutely dry soil), weed control, the introduction of large doses of organo-mineral fertilizers when ploughing grain, high labour organization during harvesting.

The bean varieties, of which there are many, can first of all be divided into codon varieties, characterised by a curly stem and therefore requiring support (vegetable gardens), and bush varieties, which have a straight and low stem and therefore do not require support (walking beans) and for this reason are of primary importance in field culture.

Beans are a crop of wide-row sowing, so in the field crop rotation it is placed in the row crop wedge. The best predecessors are a layer of perennial grasses, winter crops going on clean manured fallow, spring cereals and winter crops after occupied fallow, row crops. Beans themselves belong to the best predecessors of all crops of crop rotation.

Among other grain legumes, beans are highly responsive to the application of complete mineral fertiliser and manure. Manure at a dose of 12-15 tonnes per 1 ha, applied in autumn under the grain, is the best fertiliser for beans.

An important biological feature of beans is their high requirement for soil moisture and heat. In this regard, soil tillage should be aimed at maximising soil moisture accumulation and its economical use.

The first cultivation is carried out to a depth of 10-12 cm in 7-9 days after harrowing, the second - to the depth of seed embedment before sowing.

The quality of agricultural products is determined by the totality of external conditions, one of the main places among which belongs to soil nutrition. Its level is conditioned, in addition to the available reserves in the soil, by organic and mineral fertilisers. But the quality of agricultural products is also strongly influenced by various agrotechnical methods of cultivation of plants. Of great importance is the determination of the influence of seeding rate (plant stand density) on the physical characteristics of seeds. The most effective seeding rate that increases protein content of seeds should be considered 120 kg/ha. Depending on soil and climatic conditions, soil fertility, purpose of cultivation, variety and other conditions, bean stand density cannot be the same. Thus, according to M.S.Savitsky (1956), I.Haenena (1978) [45] and other authors, when determining seeding rates, it is necessary to take into account - species, variety and plant variety, climatic conditions, soil condition, seed quality, purpose of the crop, method and term of sowing, optimum number of stems, productive tillering, absolute seed weight, percentage of plant survival, etc., when determining seeding rates. Some authors directly propose weight seeding rates. Thus, W.H.Brau (1937), T.A.Kuesselbach, W.E.Luhes (1948), S.S.Salmon, O.R.Matheus and R.W.Lukkel (S.W.Salmon, O.R.Matheur, R.W.Luckel, 1953), O.Fuher and Stauffer (O.Fuher, W.Stauffer, 1978) indicate it for cereals within 100 kg/ha [49]. At the same time, there are known data (A.A.Kornilov, 1968; F.M.Kuperman, 1969) indicating that since the yield is the final product of photosynthesis, its value should be determined by the assimilation surface of plants (ASP), which, in turn, may depend on the feeding area under natural conditions and the level of agrotechnics [42].

Beans are sown by wide-row, nesting and square-nesting methods. In the wide-row sowing method, the width of row spacing is set 45-60 cm, in the square-nest method - 60 cm, in the nest method - 60 X 30 cm (2-3 seeds in a nest). Bush beans are sown in rows with a distance between them 30-45 cm, the distance between plants in the row 10-12 cm. The width of row spacing when sowing curly varieties is increased to 60 cm, and between plants in the row leave 7-10 cm [7;98]. Bean bean bears seed pods on the soil surface, so it does not tolerate deep seed placement well. Bean sprouts appear in a sporadic manner. If sowing is carried out at a soil temperature at the depth of seed embedding of 6-8^0 C, the sprouts appear on the 16-18th day. If sowing is carried out at a soil temperature at a depth of 5-6 cm - 15-18 ° C, then the sprouts appear on the 7-9th day. Therefore, beans are usually sown relatively late in all cultivation zones. In the foothills of the Caucasus, sowing dates are late May - early June. In the north-eastern part of Azerbaijan the earliest sowing date is 1-5 April, and in Apsheron the earliest sowing date is 15 April.

In the cotton zone, beans are sown only under irrigation, in these conditions the best sowing dates are middle and late. According to the materials of Azerbaijan State Breeding Station, in 1950 good results were obtained when sowing beans in the second decade of April.

Common bean seeds are usually sown to a depth of 3 - 5 cm, on dry loose soil a little deeper - up to 7 cm. On heavy soils it is recommended to sow seeds 3-4 cm deep. As established by D.I.Fursov (1968), with increasing the depth of sowing from 4 to 12 cm, the sowing-growing period increases by 4 days at sowing date of 28 May and by 2 days at sowing date of 29 June due to the onset of high temperatures [90].

Observance of bean sowing rates is the most important rule of agrotechnics.

Reducing the seeding rate in all cases leads to lower yields.

Bean sowing rates in field crops depend on the cultivation zone, bean species, bushiness of the bean variety and seed size. In new, more northern areas of field cultivation and in highland areas, as well as in stubble sowing, it is necessary to sow a large number of seeds (up to 400 thousand). Reduce the number of sown seeds in wide-row sowing (row spacing 60-70 cm) and when sowing on irrigated areas, as well as when sowing strongly branched varieties of lima and multiflowered beans. Seeding rate varies depending on seed size from 60 to 350 kg/ha. Small-seeded bean species are sown in the amount of 12-35 kg/ha.

This combination of late and small sowing makes rolling mandatory. However, according to Prof. Vorobyov's tests on small plots (obviously, manual cultivation), the Tepari variety gave good results when sown at 7 cm. Development at increased depth was still delayed [74].

Since seed size varies greatly, the seeding rates vary considerably: small-seeded varieties are sown in amounts not exceeding 70 kg/ha, while the rate for large-seeded varieties is brought up to 100-150 kg.

Tillage is compulsory; row spacing should be limited to 35-40 cm, given the low growth characteristic of bush varieties. Sowing can be carried out with grain drills, with top grain feeding. The plants are allowed to be densely arranged in the rows; bean bursting is never done. Repeated hoeing and hand weeding in the rows is necessary.

Sowing time and plant density have been shown to influence bean yield (Kapur, R., Satija, D.R., Dahiya, B.S., 1987; Popovic, S., Stjepanovic, M., Bosnja D., Zoric, 1988; Kolompikova, Vasyagin, Kireev, 1988; Kalaydjeva, 1989). It was found that as sowing density increased, the number of beans per plant and the weight of beans per 1 plant decreased, resulting in lower yields. Increase in row spacing width and sowing density negatively affect such indicator of crop productivity as the number of beans per plant. The highest number of beans per plant was obtained at row spacing width of 14 cm and sowing density of 25 germinated seeds per m^2 [100;46;44].

With increasing plant density and sprinkling, bean plants had higher height and higher attachment of the first bean from the bottom. It is recommended to grow beans under irrigation, wide-row sowing method, with high plant density. At the same time, selection work for high attachment of the first bean and resistance to fungal diseases was proposed.

Production is recommended to sow beans with row spacing width up to 25 cm with plant density for non-determinant varieties 173 thousand plants/ha, for determinant varieties - 220 thousand plants/ha.

Further care of bean crops consists of systematic soil loosening in the inter-row spaces, weeding of weeds in the rows and fertilising. The first cultivation is usually carried out 10-12 days after the emergence of seedlings, and subsequent cultivation (2-3 times) - depending on the degree of weeding.

For high yields, the soil most often lacks nitrogen, phosphorus and potassium. However, increasing the amount of fertiliser does not always have the expected effect.

The works of Soviet scientists (D.I.Pryanishnikov, A.I.Sokolov, M.M.Sazanov, I.V.Krasovskaya, P.A.Vlasiuk, S.P.Kulzhinsky, R.R.Khusainov, etc.) established that leguminous crops are very responsive to the application of various forms of fertilisers, including phosphorus and potassium. Their efficiency depends largely on soil conditions, as well as on combinations of these types of fertilisers with others, doses and their application [58].

D.N. Pryanishnikov, S.P. Kulzhinsky, L.I. Govarov, K.K. Alakseev, E.I. Barulina, V.S. Muratova, K.G. Prozorova, K.A. Pozdnyakov, V.S. Fedotov, R.R. Khusainov, S.S. Ilyin, M.N. Tilman.Prozorova, K.A.Pozdnyakov, V.S.Fedotov, R.R.Khusainov, S.S.Ilyin, M.N.Tilman, Y.N.Rabinovich, V.N.Beketov, I.V.Yakushkin, V.N.Chirkov, M.Ali-Zadeh and others showed that phosphorus and potassium fertilisers, applied both separately and jointly, have a positive effect on growth and development, root formation of leguminous crops and increase their yield [58].

Watering. All varieties of common bean are unstable to soil and air drought. In arid areas this crop gives low yields even on irrigated lands. Satisfactory yields are obtained only in areas with annual rainfall of at least 450-500 mm in the absence of prolonged droughts and dry winds in the phases of flowering, setting and filling of beans. At the beginning of bean growth, overwatering causes severe leaf overgrowth - to the detriment of flowering and fruit formation. Watering of seedlings can be stopped before flowering starts (except during dry periods). The plants need moisture especially during flower opening and bean formation.

The weekly watering rate is 15-20 litres of water. In 8-10 days after fertilisation of ovaries, juicy fleshy ones the size of a wheat grain are formed.

Green beans of early maturing varieties are harvested 60-70 days after the emergence of seedlings (every 5-6 days.), trying not to damage the plants. Overripe beans become fibrous.

Beans are **harvested** at leaf fall, when about 70-80 % of the beans are ripe. Low temperatures usually delay seed ripening, dry weather and high temperatures, as well as short rains interspersed with good weather, accelerate bean ripening.

Beans are harvested when most of the beans are yellow, but do not overharvest them. When harvesting beans that are too old, a significant part of the crop is lost due to bean cracking. Seed harvesting methods: direct and separate. Seeds were hatched from beans on the day of harvesting, dried in beans or on plants. It was found that the optimum harvesting date of beans comes when 80-90% of beans with yellow, greenish-yellow colour in the crop. At earlier harvesting the bean yield decreases as a result of not before the seeds are full, and at delayed harvesting - due to increased losses. Therefore, when setting the bean harvesting date, it is necessary to take into account not only the % of yellow beans, but also yellowish-green and greenish-yellow beans.

1.5. Distribution of CIS beans and the Republic of Azerbaijan

There are currently 17 known cultivated bean species, including 8 Old World and 9 New World species, which are in various stages of utilisation.

American beans became known in Europe only after the discovery of America due to the mass importation of seeds from newly conquered countries, first as rare ornamental plants in botanical gardens. From the beginning of the 17th century they became vegetable crops, from the 18th century - field crops.

Beans entered Russia by various routes. The most probable is importation through Arkhangelsk in the second half of the 16th century. In the 17th century, beans could have entered Russia through ambassadors sent to England.

Under Peter the Great, foreign gardeners sowed beans as an ornamental plant. Until the 18th century, beans were used in Russia as a vegetable and ornamental plant, and only from the middle of the 18th century. Beans were sown in the field in very small areas in the provinces of the Black Earth strip, mainly in the south-west.

The cultivated beans reached the USSR in several ways: mung bean reached the Central Asian republics through trade relations with the countries of West Asia before our era. The common bean was imported from the middle of the 16th century, but as a vegetable plant it gained some importance only from the time of Catherine II; in Georgia it became known from the middle of the 17th century.

On the former territory of the USSR, the earliest establishment of the bean crop called lobia took place in Western Georgia. According to some assumptions, the bean came here from Turkey, according to others, directly by sea from the countries of the Mediterranean basin. L.L.Dekaprelevich wrote that data on the penetration of the bean are scarce and it is difficult to imagine when exactly and from where the bean was brought to Georgia. However, there is no doubt that at first the bean got to Western Georgia and its culture gradually moved to the East [32;33]. I.A. Javakhishvili believes that beans penetrated into Georgia from Turkey in the first half of the 18th century, first to Guria and Megrelia, and then further to the eastern regions of Georgia [42].

In production conditions by the beginning of the 20th century there were only two species of beans of American origin in crops: common bean - *Phaseolus vulgaris* L. and *multifloral* bean - *Phaseolus multifloris* L. In experimental conditions, two more species of beans of American origin were tested on a very limited scale: Lima bean - Ph. *lunatus* L. and sharp-leaved or *tepary bean* - Ph. *acutifolius,* var. *latifolius.* Historical and geographical sources indicate the culture of mash under irrigation in the territory of the modern Uzbek Republic in the IX-X centuries AD. Much later in the 19th century, data on the ancient culture of mash in Khiva appeared.

In Turkestan in the late 19th century and the first quarter of the 20th century, mungbean was a common crop under irrigation, as well as on plots of rainfed land. In the Far East, mungbean was cultivated by Koreans and Chinese a few years ago.

In Georgia, mungbean has been known for a long time, but never had economic importance here. In Azerbaijan, it was cultivated in very limited areas, on farmsteads. In the late 18th and 19th centuries, mungbean was known in the lowland zone of Dagestan and on the Terek [42].

In the European part of the USSR, the main bean is the common bean. The multiflowered bean in the form of one white-seeded variety "Lopata" is cultivated on a small scale in some places in the Voronezh region and Ukraine. Lima bean is of no economic importance in the USSR, except in the Transcaucasian republics, where it is profitable. Tepari occupies limited areas in the Azov-Black Sea, Orjanikidzevsky Krai and the ASSR of the Volga Germans. Asian beans (rice bean, mung bean) are gaining some interest for sideration purposes in the Transcaucasian republics.

The Transcaucasian republics, where beans are the only legume crop, are cultivated in the lowlands and mainly in the foothill and mountainous parts up to an altitude of 1,500 metres. Bush beans, which are rarer, are found in pure crops and curly beans in mixed crops with cucurbits.

The bean crop of the USSR is expanding rapidly. For 1935 the plan of bean sowing was approved in the amount of 165 thousand hectares. In 1936, the accounting of actual sowing gave a value of 187 thousand hectares for the USSR. More than a half of this area (103.8 thousand ha.) belongs to the share of the Ukrainian SSR.

Apart from beans themselves, the national statistics also highlights lobia, which was sown in Georgia on 5.5 thousand hectares.

The bean crop is relatively young for Russia. Systematic breeding work with vegetable beans began in the 20s and was carried out both in the countries of the present near abroad and in Russia. The first works on bean breeding were started in 1934 by A.V.Krylov; materials on beans were received from Gribovskaya experimental station. For incomplete 10 years he achieved significant results: already in 1943 the varieties of vegetable bean Golden Mountain, Triumph sugar bean, Saksa without fibre 615 were released. In 1949, researcher N. Michelman continued the theme of breeding new and improving existing varieties of beans, development of schemes and sowing period of vegetable beans [42].

In the southern regions of the USSR, new bean species that had not been cultivated before were positively tested, for example: rice bean (Ph.calcaratus Roxb.), grown for fodder and green fertiliser; Ph.semierectus L. and Ph.aconitifolius Jacq., which have fodder value. Ph.aconitifolius in Sukhimi has justified itself as a plant for lawns and sodding of steep slopes. Of greatest practical importance is rice bean for sowing in tea and tung plantations, as well as in a number of areas where tobacco, essential oil and other southern crops are cultivated. In addition to Georgia, Azerbaijan and Uzbek SSR, long-term experiments with different bean species were conducted in Moldova [42].

In the former USSR, beans were cultivated in the North Caucasus, the Central Black Earth regions, the steppe and north-western zone of Ukraine, the left bank of Ukraine and Transcaucasia. In 1930, the area sown with beans totalled 65.3 thousand hectares. During the war years bean production decreased;

the area under this crop in 1945 was 198.6 thousand ha.

In the Ukrainian Republic, the largest areas of bean crops are concentrated in Odessa, Dnipropetrovsk, Zakarpattya, Zaporizhzhya and Khmelnitsky oblasts.

The area under bean crops from the total sowing of grain legumes in the USSR as a whole for 1956 is 4%, and the role of beans in the public crops of collective farms is only 2%. Undoubtedly, the new purchase prices set in 1958 will favour the expansion of bean crops.

Beans as a field and vegetable crop before the Great October Socialist Revolution were cultivated only in the southern part of Russia.

The development of bean culture in Azerbaijan dates back to the second half of the 18th century; here beans were cultivated in a small variety and had only consumer value as a vegetable garden crop. The ways of distribution of Asian and American bean species in Azerbaijan are given [42].

In the new tea-growing areas of the Azerbaijan SSR (Lenkoran, Astara), high-growing bush varieties of lima bean are successfully used to shade young tea plants in spring and in the first half of summer. In cotton-growing areas of Central Asian republics, Kazakhstan and Azerbaijan, in addition to its high nutritional qualities, mungbean is also valued as a good precursor of cotton.

N.R.Ivanov (1961) on the basis of botanical and geographical study of common bean varieties of the world collection of VIR-a, 30 ecotypes were established according to the character of growth, duration of vegetation, degree of branching form, biology of flowering. Five ecotypes are widespread in the CIS: 1) northern local, 2) forest-steppe, 3) steppe, 4) Caucasian, 5) Carpathian. Caucasian ecotype are curly (to varying degrees), medium- and late-ripening; they are characterised by significant foliage and increased demand for moisture and heat. They are often sown in mixed crops with maize. Among the released varieties of this ecotype are: Galibiat local, Red Indian local, Ramazanskaya, Chitis quertskha and a number of local varieties of Gurzinskaya SSR [42].

Galibiat local (Great Northern, Great Northern, or White imported). It is distributed (since 1948) in Azerbaijan SSR, except Nakhichevan ASSR. Shape of the bush is slightly curly. Leaves are green, leaflets broad, flowers white. Beans are flat, sabre-shaped, 8-11 cm long, husky; immature beans are green. Change kidney-shaped, white, weight of 1000 seeds 260-300g. Late maturing; vegetation period 80-120 days. Affected by fungal diseases and bacterioses.

Piada local (Tsanava type). Local variety of the Azerbaijan SSR. It was zoned (1951) in the Azerbaijan SSR, except for Nakhichevan ASSR. The plant is bush-shaped, 25-40 cm high, medium branched. Leaves weakly corrugated, green; leaflets broadly ovate, acuminate. The flowers are purple. Immature beans are green, then red-mottled; mature beans are pale reddish-brown; straight, distinct, 8-10 cm long. The parchment layer is formed late. Seeds are oval-shaped, large; mottled colour, the main background pale yellow-pink, with light red spots; weight of 1000 seeds 400-450 g. Soon-ripening, relatively resistant against bacterioses. Shifts are well digested.

Red Indian local. An old variety of the Georgian SSR. It is distributed (1946) in the Georgian SSR (without Abkhazian and Adjarian ASSR and South Ossetian Autonomous Oblast) and Azerbaijan SSR (without Nakhichevan ASSR).

The most important bean varieties are recognised as: 1) for the Karabakh regions Red Indian local, 2) for the unchanged and irrigated regions of the Republic the variety Galibiat, 3) for the temperate regions of the Republic the variety Yerli Piada.

At present, a great variety of bean varieties, forms and ecotypes can be found in our country (Figure 1.5). In addition to common bean, Lima bean and multiflowered bean grow here, which have both food and ornamental value.

Figure 1.5. Seed diversity diversity of bean forms and ecotypes.

Despite all the advantages of beans, they are one of the least studied crops in Azerbaijan. One of the reasons of insignificant sowing of beans in production is the lack of varieties meeting the requirements of modern farming in the republic. On the diversity of bean forms, which are excellent source material for breeding new varieties, but its selection until recently very little engaged in [6]. True created a few varieties, such as AzNIZ-352, Zulal, Sevinj, Isa, etc., but not enough.

Among grain legumes in Azerbaijan, beans are the most valuable food crop. It is characterised by a high content of high-grade and at the same time cheap food protein (up to 33%) with a relatively well-balanced amino acid composition. Increasing bean production will undoubtedly help to improve the level and quality of protein nutrition of the population. On the territory of the Republic of Azerbaijan cultivated varieties of common bean are mainly Caucasian ecotypes.

Beans, originally an imported plant for Azerbaijan, have taken a place in the country's cuisine (diet) over time. Throughout the year a variety of bean dishes are prepared from cold snacks to hot dishes. It should also be noted that due to the fact that beans are grown mainly in homestead plots, the population's demand for this product is not satisfied by local production. Most often beans are brought from abroad (beans are imported mainly), which causes additional costs. Taking all this into account, it is necessary to expand bean planting areas to meet domestic demand. In turn, this requires the development of fertile and mechanically harvestable varieties.

In spite of favourable growing conditions, beans in Azerbaijan have not received a great spread, especially varieties of food direction. Therefore, we decided to study different bean varieties of the main biological, morphological and economically valuable features. In this regard, it is very relevant to conduct an in-depth study of the variety of samples of the VIR collection, local and introduced varieties in order to select the source material for breeding new high-yielding varieties of beans of intensive type, meeting modern production requirements, with good seed quality, resistant to diseases and suitable for mechanised cultivation.

In the current millennium and warming, which is also observed in Azerbaijan. At the same time, the created varieties must be early maturing and withstand elevated temperatures. Scientists have calculated that climate change by 2050 may halve the amount of land suitable for growing beans.

Breeders at the International Centre for Tropical Agriculture (CiAT) in

Cali, Colombia, are trying to find a solution to this problem. Using data from genetic banks, they have identified about 30 varieties of common bean *(Phaseolus vulgaris* L.) that can tolerate elevated temperatures [95].

To do this, a team of scientists led by plant breeder Stephen Beebe, apparently particularly concerned about rising temperatures, studied more than 1,000 bean varieties. Other researchers had previously suggested all of these varieties as being able to cope with drought and grow in infertile soils. Often scientists have crossed common bean varieties (black Premo bean, red kidney bean and pinto bean) with a related species, the pointed bean or tepary bean. The tepary bean is common in northern Mexico and the south-western United States and is the most popular crop in agriculture because of its vulnerability to soil infections and small bean size, but it can survive drought and heat waves.

As a rule, bean yields begin to decline if night temperatures exceed 18-19^0 C. It is at night that bean flowers open for pollination, but excessive heat prevents this process. It is at night that the bean flowers open for pollination, but excessive heat prevents this process. The heat also causes the plants to grow more slowly.

Beeb and his colleagues have planted samples of thousands of varieties in duvvu of very hot regions whose nighttime temperatures never fall below 23^0 C degrees; in lowlands along Colombia's northern coast and in research fields at the University of Tolima southwest of Bogotá.

The researchers then examined the seed pods to see if pollination was successful. Only 30 of the varieties were the result of a cross between the common bean and the spiky bean. This suggests that some DNA traits in these varieties may help the plant to survive the heat.

Transferring the unique hardiness of tepary to modified beans is not easy, but researchers seem to have made progress. The viability of the six best varieties exceeded 64% (compared to 20% of heat-sensitive varieties).

Resistance to high temperatures is a very important trait for legumes grown in Central America and the Caribbean. Breeding new bean varieties can help farmers cope with the food crisis expected as a result of climate change. Large-scale testing of new varieties may take several years, but one of them (SEN 52, drought tolerant) has been grown and marketed since 2013 [95].

Currently, the actual problem of bean breeding in Azerbaijan is the creation of the most early maturing and medium maturing, high-yielding, determinant type of growth (height 50-70 cm), varieties with high technological properties (height of lower bean attachment 15...25 cm). Breeding work with vegetable beans is aimed at the creation of bush varieties intended for sale in fresh form and for processing (canning, freezing) with the expectation of the desired calibration of fruits and seeds. Special attention is paid to high bean quality. The properties required by processors and consumers must be combined with a complex of economically valuable traits, for which it is important to have a variety of source material.

In this regard, the aim of our research was to study the source material, identify sources of economically valuable traits, select parental pairs and obtain a series of recombinants, the parameters of which would be most suitable for the model of new generation varieties.

CHAPTER 2

CONDITION, MATERIAL AND METHODOLOGY

The research was conducted in 2005-2015 at the Apsheron base of the Institute of Genetic Resources of the NAS of Azerbaijan. Experiments were conducted at the Apsheron Research and Experimental Base (ANEB), which is located on the Apsheron Peninsula. The climate of Apsheron is dry subtropical. Summer is dry hot, autumn is warm sunny, winter is mild, almost snowless. The average annual temperature is 13.5-14.5°. Frosts are rare in winter, in summer the air temperature reaches 38-4°, and after 2010 the air temperature is high 40-45°. Average annual precipitation is insignificant 120-150 mm, relative humidity is 70.6 per cent. Summers are almost always dry. The soil is sandy loam, very poor. The climate is strongly influenced by the Caspian Sea and the semi-desert plains adjacent to the peninsula. The experimental plot is located at an altitude of 80 m above sea level.

The maximum shade temperature in July and August sometimes reaches 44°C. The driest months are July and August, precipitation falls mostly in the winter-spring period. During the vegetation period of beans spring sowing dates, precipitation falling here does not ensure their normal development and, therefore, crops need artificial watering at least 4-6 times.

The research covered more than 130 samples of local, domestic and introduced bean collection. They belong to 3 species: *Phaseolus vulgaris* L. - common bean, *Phaseolus coccuneus* L. - fire or multiflowered bean, *Phaseolus lunatus* L. - lima bean.

Seeds were sown 5X5 m in plots, row spacing 45-60 cm. The study of collection material for a number of morphological and economically valuable traits (sex counts, observations, analyses) and evaluation was carried out according to "Methodological guidelines for the study of grain legumes and beans". Selection of sheaf material for analyses on the elements of yield structure was carried out manually as the varieties matured.

Description of bean varieties and samples on morphological traits was carried out in several stages: in the period of sprouting, in the period of flowering, in the period described by the following traits: type and shape of the bush, colouring of the stem and leaves, shape, pubescence, surface of leaves, colouring of flower parts, shape, colouring, presence of fibre and parchment layer in the pod, colouring and shape of seeds, pattern and colouring of seed pattern.

The collection contains bean varieties with different types of bush: bushy, semi-bushy, curling, curly; bush form: erect, spreading, stalked. The colour of the stem varies from light green to purple. Leaves of the presented varieties differ in colour: light green, green, dark green, in the shape of the top: retracted, pointed, blunt. The colour of flowers of collection specimens is represented by the following shades: white, light pink, pink, purple. The beans are divided by colour into yellow, green and purple. Seed shape is elongated, oval, elliptical, semi-convex, rounded. The following colours are present in seed colouring: white, cream, pink, beige, red, purple. Seed pattern is unicoloured and bicoloured.

In spite of favourable growing conditions, beans in Azerbaijan have not received a great spread, especially varieties of food direction. Therefore, we decided to study the following main issues under different cultivation conditions of bean varieties in Absheron.

1. Biological, morphological and economically valuable traits of the studied bean varieties.
2. Influence of different agronomic practices on the degree of fruit formation and yield structure. In this direction; sowing methods, sowing dates, depth of seed placement were studied.
3. Selection of the best varieties of beans of food direction for introduction of production of the republic.

CHAPTER 3

3. Characterisation of bean samples on economically valuable features

3.1. Morphobiological features of grain bean samples as a source material for breeding

3.1.1. Duration of the growing season. It is of interest for breeding to know the amplitude of variability of the growing season in relation to certain varieties and forms. In this case, it is important to study the growing season not only in total, but also for individual phases of growth and development. It is known that two genotypes with the same vegetation period can differ significantly in the duration of interphase periods. These differences are important for breeders. In the process of phenological observations, differences between varieties and samples are studied in terms of the length of the growing season, the duration of individual phenophases, which make up the entire period of plant development from seed to seed. Based on these systematic observations, the sources of early maturity are identified, pairs are selected for crossing, and the source material is divided into early maturing, medium-early, medium-late, medium-late and late maturing groups.

It should be noted that the duration of the sowing - sprouting period does not depend on the varietal characteristics of beans. Duration of this interphase period depends on the depth of seed placement. D.I.Fursov (1968), noted that when seeds were sown to a depth of 4 cm, this period was equal to 10 days, 8 cm-12 days, 12 cm-14 days, 16 cm-15 days [90].

The length of the growing season is determined both by the hereditary nature of the plant organism and the totality of its growing conditions [16;64;37;12;43;65].

It is known that of all environmental factors, air temperature and precipitation have the greatest influence on phenology [64].

A.A.Abdullaeva (1979) considers the length of the growing season and the interphase period from sowing to flowering as complex polygenic traits [2].

The growing season largely determines the suitability of a variety for cultivation in a particular area. Many economic and biological traits and properties of a variety (resistance to drought, diseases and pests, product quality, etc.) and ultimately its yield are related to the duration of the growing season.

When selecting beans for early maturity, it is necessary to take into account the duration of not only the entire growing season, but also the interphase stages: sprouting - flowering, sprouting - biological maturity, and for vegetable varieties: sprouting - flowering, sprouting - technical maturity, sprouting - biological maturity.

Detailed study of these periods in different varieties under specific conditions is of certain scientific and practical interest for breeding. The duration of the growing season depends on varietal characteristics and growing conditions. Such conditions as soil and air temperature light [55;18;28;29] have the greatest influence on the variation of the growing season.

Three periods are distinguished in beans:

1) From sowing to sprouting;
2) From seedling emergence to flowering;
3) From flowering to seed ripening on the plant.

According to our observations the duration of sowing - sprouting depending on meteorological conditions varies within quite wide limits, from 9 to 24 days. Thus, in order for the bean plants to germinate and give friendly sprouts, we need the sum of average soil temperatures above 12^0 C, equal to 166-183^0 C, and sufficient soil moisture. It should be noted that in our experiments the duration of the sowing - sprouting period did not depend on the varietal characteristics of beans. However, the duration of this interphase period depends on the depth of seed placement. Thus, when seeds are sown at a depth of 4 cm, this period is 10-12 days, at 10 cm - 12-15 days [6].

The length of the day is crucial for beans during the germination and flowering period. In short-day bean varieties, as we move northwards, the sprouting - flowering period increases, while in long-day bean varieties, on the contrary, it decreases. This was shown by A.V. Doroshchenko and V.I. Razumov on experimental crops [7;76]. According to N.R.Ivanov in geographical experiments of VIR (1952-1954), a significant variability of the duration of the interphase period from sprouting to flowering and depending on the year and place of cultivation was observed [42]. .

The duration of the flowering - maturing period of beans depends on flowering, air temperature and humidity during this period.

According to A.I.Rudenko and M.N.Mamedov the vegetation period of beans of different varieties is shorter, the higher the average air temperature for the period [81;57].

The duration of the next period - from sprouting to flowering - depends on the varietal characteristics of the bean; depending on the variety, the variation in the duration of this period is from 2 to 5 days. At this time the bean is less demanding to moisture, more resistant to drought, as the whole root system of it develops more intensively than the stem and all its equal or greater weight of the stem. Later on, this ratio decreases.

The length of time from flowering to maturity depends on the availability of heat and moisture. During rainy periods, bean maturation is delayed. The duration of the flowering to ripening phase is more variable than that of the seedling to flowering phase.

As a result of our research, it was found that the duration of the growing season as a whole and individual interphase periods varies significantly depending on sowing dates, cultivation conditions and biological characteristics of varieties.

According to our phenological observations, the period from the beginning of flowering to seed ripening in early maturing varieties runs from the beginning of June to the second five days of July.

Bean species are very diverse in terms of the length of the vegetation period: from early maturing, with a vegetation period of 70-75 days, to late maturing, with a vegetation period of more than 200 days. For soon-to-ripen forms the sum of active temperatures during the vegetation period is 1500°; medium-ripening forms - 1500°-2000^0 ; late-ripening forms - 2500-3000° and even 4OOO^0 C. Bean seeds of different varieties and species need different temperatures for germination. For example, common bean requires 8-10^0 C - at this minimum temperature sprouts appear on the 20-25th day after sowing. Dark-coloured seeds of various bean species usually start to germinate at temperatures 2-3^0 C lower. The optimum temperature during budding and flowering periods in beans is 22-28^0 C [7].

The length of the growing season of common bean depends on the variety, weather conditions and geographical latitude of the area. For the same variety, the higher the temperature in the area, the shorter the growing season.

The duration of the vegetation period and its constituent interphase periods in our experiment depended on both varietal characteristics and weather conditions. As a result of research, in accordance with climatic conditions, all studied samples were grouped into 4 groups of ripeness according to seed maturation:

early maturing (58-70 days) - 29 samples, medium maturing (70-80 days) - 44 samples, medium-late (80-90 days) - 6 samples, late maturing (90-120 days) - 6 samples. The shortest interphase period between sprouting and flowering compared to the standard variety was observed in samples AzePHA-20, 42/1, k-15275, Zulal.

To increase bean acreage, not only high-yielding varieties are needed, but also early maturing varieties. A short growing season solves many problems: care from early and late frosts, drought, disease and insect damage. The task of obtaining early maturing varieties (the vegetation period of which does not exceed 80 days) is the most important in the world breeding [51].

In our researches over the years of study in this direction, the sources of early maturity were identified: Yerli Piada, AzePHA-20, k-15274, t/3, AFGO-22, t/37, AzePHA-14, AFGO-27 (Table 3.1).

Characterisation of sources of early maturity in beans on average for 2008-2014. Table 3.1.

Variety	Duration, days			
	sprout-blooming	beginning of flowering-end of flowering	blooming ripeness	growing season
Yerli Piada	32	6	33	65
AzePHA-20	32	6	34	66
к-15274	35	5	36	71
т/3	29	5	39	68
AFGO-22	30	12	33	63
т/37	33	5	37	70
AzePHA-14	26	7	42	68
AFGO-27	32		32	64

Rearing rate is a stable trait. The coefficient of variation of this trait does not exceed a few per cent. However, it is an extremely important indicator. A shorter growing season has a significant impact on yield, quality and range expansion.

We found samples of common bean with the same duration of the full vegetation period, but significantly differing in the duration of interphase periods. Comparative study of such samples has shown that at the same duration of vegetation period, the forms with a short period from flowering to maturity are distinguished by relatively high productivity. This is especially important in the selection of parental pairs for breeding early maturing high-yielding varieties of common bean, "...the vegetation period is a capital section of breeding, for it is inseparably connected with many traits..." (Vavilov, 1965). (Vavilov, 1965). The duration of the growing season is a complex quantitative trait that depends on the variety, climatic conditions of its cultivation, latitude and altitude, as well as many other components [49;102].

For breeders, not only the growing season but also the organisation of its components is important [52].

The ways of obtaining early maturing high-yielding forms are different. Some authors argue that it is possible to overcome the undesirable relationship between early maturity and productivity decrease by increasing the indices of the least variable traits. These include 1000 seed weight [27;68] and the number of seeds in the pod [89;20]. Others believe that the creation of early maturing high-yielding forms is possible only with a balanced increase in all elements of productivity: the number of productive nodes, beans on the peduncle, seeds in the pod and 1000-seed weight [91;22].

When selecting beans for early maturity, it is advisable to use forms from the group with the lowest variability of individual phases of the growing season [4].

Duration and variability of the growing season and its component phases in common bean under Apsheron conditions (2010-2014).

Maturity groups	Sprouting-flowering	Flowering-ripening	Sprouting-maturing
----------------			--------------------------
Early maturing	26-33	33-36	62-69
Mid-maturing	30-36	37-40	67-76
Mid-late	31-45	51-53	82-98

In our studies, the least variability of vegetation period and its separate phases was observed in samples of early-ripening and medium-ripening group. Distribution of different bean species by ripeness.

Distribution of varieties and forms of common, lima, and multiflower beans by ripeness under Apsheron conditions (2010-2012).

Early maturing 58-70	medium maturing 71-80	medium-late 83-90	late maturing 92-120
	Common bush bean		
16	35	3	1

		Semi-high and curly.		
	15	34	13	9
		Lima beans	-	2
	-	Multiflower bean	-	2
Totals: 130	23,8 %	53,1 %	12,4%	10,7 %

In addition, AzePHA-20, AzePHA-3, AFGO-27, which are of interest in the development of early maturing varieties intended for one-time mechanised harvesting, have been identified.

3.1.2. Morphological traits. The studied samples differed in a whole complex of traits. Diversity was observed in size, shape and colour of seeds and flowers; size, shape and colour of immature beans; leaf size and shape.

According to literature data, correlations of traits showed that the size of primordial leaves (length, width) have strong positive relationships with the size of true leaves, length and weight of bean in the phase of technical ripeness, size and weight of 1000 seeds. The strong positive correlation of primordial leaf size with the mass of 1000 seeds is explained by the fact that it is an indirect dependence, because the bean size depends on the mass of 1000 seeds (seed size). The correlations we have identified are of interest because they *allow* us *to significantly accelerate the breeding process.* Knowing the nature of the relationship of traits, in particular seedlings in the phase of 2-4 true leaves.

Schematically, the dependencies would be as follows: large primordial leaves ^large (broad) mid leaves "-▶ large bean <-> large seeds, or vice versa: Small primordial leaves <-> small (narrow) mid leaves <-> small bean <-> small seeds.

A complex evaluation of phenotypic variability of varieties was carried out for the following traits: plant height; height of lower bean attachment; number of productive nodes; beans; seeds per plant; number of beans per one productive node; seeds in bean; seed weight per plant and 1000 seed weight.

Plant height - responsible for plant growth character, determined by the plant genotype, growth and development conditions, and also influences the improvement of harvestability in beans. The length of the plant stem ranges from very short (20 cm) to very long (250 cm). Beans are divided into two types of growth pattern: indeterminant (incomplete) and determinant (complete). We studied varieties had determinant and non-determinant type, plant height varied from 25 to 180 cm. According to this characteristic, the samples were divided into three groups: for determinant with low (less than 35 cm), medium (35-50 cm) and high bush (more than 55 cm), and for nondeterminant with low (45-60 cm); medium (60-85 cm) and high (85-110 cm and more). According to the results of the research, 26 short-stemmed, 30 medium-stemmed, and 18 long-stemmed cultivars were included in the short-stemmed group. Medium-stemmed samples are the most suitable for creating an optimal model of a new generation variety.

Distribution of varieties and forms of common bean by plant height when grown under Apsheron conditions (2005-2010).

	short-stemmed		medium-long-stemmed	Long-stemmed
	20-45 cm;	50-75cm;	80-120 cm;	18
Common bush bean				
	42	13	-	-
semi-hung				
	-	15	48	8
Lima beans				
	2	-	-	-
Multiflower bean				
	-	-	-	2
Totals:130	55,3%	21,5%	36,9%	7,6%

Our studies showed that the plant height of the tested samples varies year by year and depends mainly on the biological characteristics of the tested sample and meteorological conditions of gad (Figure 1).

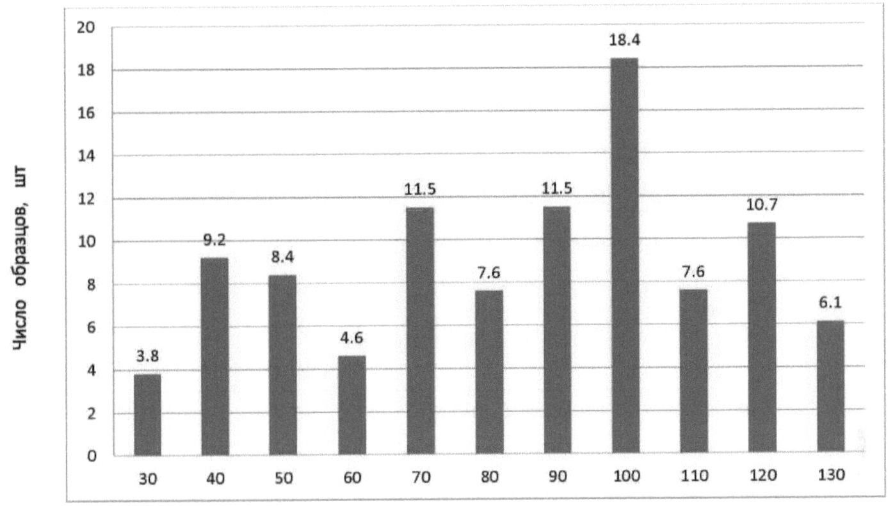

plant height, cm

Figure. 1. Distribution of bean varieties of the collection nursery by plant height.

The height of lower bean **attachment** is a decisive, technological trait that reduces bean seed losses during direct harvester harvesting. Technological ability, along with such important traits as productivity and early maturity, is the main indicator of a variety's suitability for mechanised cultivation and harvesting. Processability is a complex trait that includes the following: variety morphotype, plant height, lower bean attachment height, non-legibility of plants, as well as bean non-cracking and seed non-sprouting. Lower attachment leads to yield loss at harvest and higher attachment leads to underharvest of biological yield [26;71;25]. It is conditioned by the total number of internodes on the plant, their distance (especially the 4th and 5th, as the first productive nodes are laid at their level) between the root neck and beak of the bean. The average value of the trait varied from 10.5 cm to 28.5 cm. Low attachment was characterised by varieties with a strong, spreading bush, and high attachment was characterised by varieties with a compact, erect bush. According to the parameters of variety suitability for mechanised harvesting, the samples were divided into two groups: those with low lower bean attachment up to 15 cm and those with high attachment above 15 cm. When studying the height of lower bean attachment, A.K. Leshchenko [51] and V.I. Sichkar noted that this indicator is mainly determined by environmental factors and 33% is determined by the genetic characteristics of the variety.

Distribution of varieties and form of common bean height of lower bean attachment

	Low attachment 10.5-15.0 cm	High attachment 15.1- 30.5 cm
bush		
	38	17
Half-hung and half-hung		
	33	24
Totals: 112	63,4%	31,6%

Twenty-four accessions were characterised by high attachment. For further work, *the sources of high compact bush and high attachment of lower beans were selected.*

Bean traits are particularly important in bean breeding, as they are the main indicators characterising the variety. Thus, external marketable qualities of immature beans are determined by fleshiness, absence of parchment layer in the flaps and fibres in the seams, intensity and uniformity of colour, shine, as well as their smoothness and evenness.

Bean shape. The varieties with rounded beans are of greatest interest to the canning industry. They are characterised by slow overripening and high nutritional values.

Bean **length** is a characteristic varietal trait that depends to some extent on weather conditions and determines marketability and canning behaviour, as beans are canned with whole beans or in sliced form.

In our studies, the length of beans across the collection ranged from 6.1 cm to 18.0 cm. As a result of the study, the samples were divided into varieties with short (6.1 - 10.0 cm) - 38 samples, medium (10.1 - 14.0 cm) - 30 samples and long (more than 14.1 cm) - 4 samples.

Distribution of varieties and forms of common bean by bean length

Short	average	long

	7.6 cm-10.0 cm	10.1-14.0 cm.	14.1-18.0 cm.
	bush		
	18	37	2
	Half-hung and half-hung		
	20	36	2
Totals:			
115	33,1%	63,5%	3,4%

Bean crackability is an important breeding trait for mechanised cultivation and harvesting of bean varieties is stover.

3.2.1. Productivity and elements of its structure

One of the most important tasks facing the agriculture of our republic is still to increase grain production. The solution of this problem will largely depend on the success of breeders' work on the creation of high-yielding varieties.

Modern breeding should have mobility in solving actual practical problems, for which it is necessary to have a variety of source material, which can be a source of both individual traits and their various combinations. There is evidence that the analysis of yield structure by its constituent elements makes it possible to determine the leading importance of individual ones in the formation of grain level in certain zones and under conditions of different backgrounds. S. Boroevich (1984) refers the number of plants per unit area, the number of grains per plant and the mass of 1000 grains to the main components of grain yield [12]. Grain yield is usually ensured by maintaining equilibrium between the components with the condition that when the value of one component increases, the value of another component decreases and vice versa. The maximum yield is the realisation of the most favourable equilibrium between all its components.

Seed productivity consists of the following elements: number of productive or fertile nodes, number of flowers (beans) on the peduncle, number of seeds in the pod, weight of 1000 seeds. Seed yield, apart from the elements of productivity, includes the number of plants per unit area by the time of harvesting.

Plant productivity in terms of green mass depends on the length of the main stem, coarseness of vegetative parts, conditioned agroecological seed productivity genetically investigated with this or that completeness.

According to V.S.Fedotov (1960), the elements of seed productivity structure can be arranged in the following order by the degree of variability coefficient: 1) bean fulfilment (number of seeds in a bean); 2) seed size; 3) number of beans on the fruit-bearing plant; 4) number of fertile nodes per 1 plant; 5) number of beans per 1 plant; 6) number of seeds per 1 plant; 7) seed weight per 1 plant [89].

Seed productivity is determined by the weight of seeds per plant, so this indicator is the most important economically valuable trait. According to N. Korsakov, the number of beans on one plant is also a relative indicator in characterising its productivity and is applied within samples with the same seed size, as the mass of 1000 seeds and the number of seeds in a bean have a wide range of variation.

According to L.L. Dekaprelevich [34] (1965), bean seed yield varies depending on soil and climatic zones of its cultivation. In the central regions of the European territory of the USSR bean seed yield ranges from 10.0 to 14.6 c/ha. In the central chernozem zone it ranges from 9.2 to 17.7 c/ha. In the North Caucasus bean seed yield is at the level of 10.6-18.5 c/ha. In the forest-steppe regions of Ukraine bean yields vary from 8.8 to 17.9 c/ha.

Productivity of vegetable bean varieties is a complex trait. It is determined by the number of beans per plant, the number of seeds in a bean and the weight of 1000 seeds, as well as the weight of beans per plant, which depends on its components: the number of beans at technical ripeness and the weight of one bean.

The number of beans per plant is a variable trait and depends on varietal characteristics and weather conditions during the growing season. In this case, it should be replaced that the number of beans per 1 plant depends on two components - the number of fruiting nodes and the number of beans per fertile node; the number of seeds per 1 plant is a derivative of the number of beans on the peduncle and the number of seeds in the pod; the weight of seeds per plant is determined by the number of fertile nodes, the number of beans per fertile node and the weight of 1000 seeds. In our experiment, the number of beans per plant by year in the studied samples varied 2008 - 2015 from 5.2 to 39.4 pieces. According to the result of analysis of variance, varieties with high (more than - 21.5) - 11 samples, medium (11.5 - 21.5) - 18 samples and low (less than - 11.5) number of beans - 10 samples were identified.

-----------------	Low	Average	--------------------------
	5,2-10,1	11.5-20.1 cm.	

			High 21,1-39,4
------------------	bush 25	24	-------------------------- 8
	semi-hung	3520	3
Totals: 115	52,2%	38,3%	9,5%

Sources of high zecjio beans - AzePHA-27; AzePHA-21; AzePHA-25; AzePHA-34; T/27 - have been identified for further breeding work.

According to many researchers, these two traits are highly correlated and are very important components that determine the level of yield per unit area [12].

Seed weight per plant characterises the seed productivity of beans, which is especially important in seed production of this crop. Seed productivity is determined by the weight of seeds from one plant, so this trait is the most important economically valuable trait. The trait is characterised by a high degree of variability and is conditioned by varietal peculiarities and weather conditions. According to most authors, plant productivity depends on the presence of favourable factors in the period "sprouting - flowering", feeding area, weight of 1000 seeds, number of seeds in a pod, plant height, oblateness. Seed weight per plant is a complex trait, which depends on weather and agro-ecological conditions. It was found that the role of external factor in the formation of this trait is 71-78%, against 19-27% of varietal characteristics.

The data of mathematical processing show that seed weight per plant has a close positive relationship for all ripeness groups only with the number of beans on the plant.

The number of seeds per plant is a variable trait and is determined by varietal characteristics and weather conditions during the growing season. The beginning of wide application of the concept of seed productivity of plants was laid by the research of T.A.Rabotnov (1945) [75], in which much attention was paid to the methodology of determining the fertility of plants, the concept of average seed productivity was introduced, i.e. the average number of seeds per individual, which can serve as an indicator of productivity of an individual.

According to the research of N.M.Verbishchky, in Rostov region, on the average for collection samples of pea the smallest coefficient of variation is characteristic for seed size, further on the degree of increase of this indicator are the following features: fulfilment, or the number of seeds in a pod, the number of fertile nodes, the number of beans and seeds per 1 plant. In our experiment, the number of seeds per plant varied from 14 to 150 pieces. Five accessions with *relatively high and stable number of seeds per plant were* identified for further work*: -/6',* k-34; k-15274; T/27; Yerli Piada. T/27 had relatively the highest number of seeds per plant at 150 seeds per plant. Distribution of varieties and forms of common bean by number of seeds per plant

	low 12-25 pc.	averages 26-39 pc.	high 40-66 pc.
		bush	
	22	30	3
		semi-hung	
	27	33	-
Totals: 115	42,6%	54,8%	2,6%

The traits seed number in pods depended on variety and to a large extent on weather conditions. According to literature data, the number of seeds in the pod belongs to weakly or moderately variable traits. Pod fulfilment, according to V.S.Fedetov, is the most stable element of yield structure. We can agree with this only with a big reservation. Varieties belonging to different agroecological groups differ significantly in the number of seeds in the pod. When selecting source material for selection, it is necessary to pay attention not only to the average fulfilment of the bean per 1 plant, but also to the maximum number of seeds in the bean. This trait is hereditary and depends on the number of buds laid in the ovary of seeds, as well as on the ability of the variety to realise this possibility. It is an indicator of the potential number of seeds produced under the most favourable developmental conditions. In determining seed productivity buffer role can be played by elements characterised by the most indicators of the coefficient of variation, for example, the number of fertile nodes per 1 plant. According to V.V. Khangildin, buffering is carried out by changing the number of seeds in a pod and the number of fertile nodes [91].

In our studies, this indicator varied from 3 to 8 seeds on average in the collection. The samples - t/15; t/7; k-14360; AzePHA-41; Yerli-2; Zulal had a high number of seeds (6-8). The source of high number of seeds in the pod - t/15 - was identified.

The **weight of 1000 seeds** is an important economic trait subject to variability and determined by varietal characteristics and weather conditions. According to A.I. Nosatovsky [64] (1956), high temperatures during the period of grain filling lead to a decrease in the weight of 1000 grains. Rainy and damp weather also leads to a decrease in the weight of 1000 grains. The weight of 1000 seeds, averaged across the collection, ranged from 170 to 624 g. Fulfilment and grain coarseness are usually characterised by the weight of 1000 seeds. One researcher believes that this trait strongly influences grain yield [54;50].

Distribution of bean cultivars and moulds by 1000 grain weight value (2010-2014).

Studied Assortment	Number of samples by weight of 1000 grains		
	shallow 169-200 г	average 205-400 г	large 408-624 г
Common bean:			
bush	9	39	6
Half-hung and half-hung	7	48	17
Lima beans		2	
Beans multiflowered			2(91.5g).
Totals: 130	12,3(%)	68,5%	19,2%

Among the bushy specimens, a greater number of them have medium seeds, while the semi-bushy and tufted specimens have medium and large seeds. Specimens with very small seeds - 4 specimens, medium seeds (205-400 g) - 46 specimens and large seeds - 20 specimens were identified. The *sources of very small white seeds* - T/27 (117.0 g), the *source of large coloured seeds* - AzePHA-3 (624 g) and *ArupnA AZIIPH seeds* AzePHA-29 (540.0 g) were identified.

Our studies have shown that the weight of 1000 seeds of tested samples varies by year and depends mainly on the biological characteristics of the tested sample and meteorological conditions of the year.

Mainly the studied samples had small and medium seeds (169-400 g). Significant variation of 1000 seed weight by year was observed (Figure 2). Large variability in seed size was found, especially in large-seeded forms, which causes instability of yields of such varieties under unstable moisture conditions.

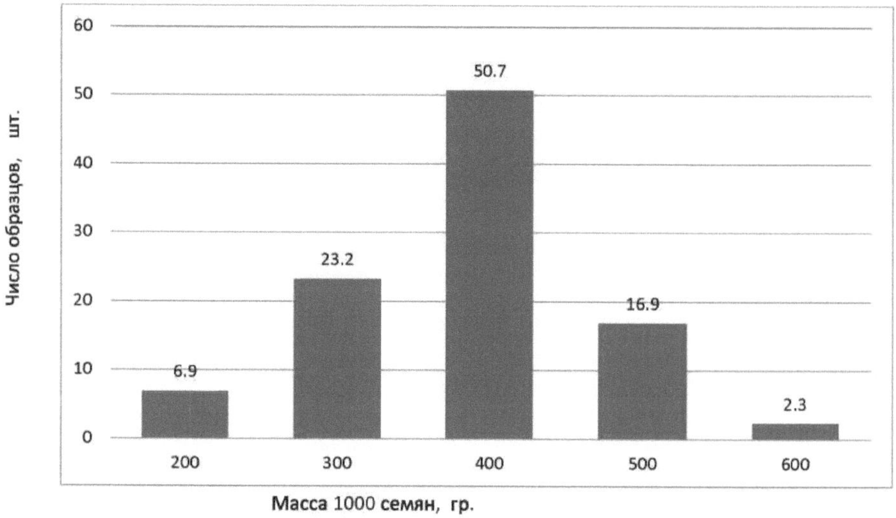

Figure 2: Distribution of seed bean collection samples (2010-2014).

Thus, among the tested 130 bean varieties and lines, a number of accessions with high 1000 seed weight stand out, which allows their inclusion in the working bean collection.

3.2.2. Correlations of productivity elements. In breeding practice it is necessary to take into account correlation dependencies, as selection on one of the traits usually directly or indirectly leads to changes in other traits. The correlations of traits differed both in magnitude and in the nature of the relationship. The level of manifestation of correlations was variable and depended on the year of study. Summarising the data of correlation analysis, the following regularities were revealed:

- high positive relationships exist between the traits determining bean productivity in biological ripeness: number of seeds and number of beans on the plant (r= 0.72), number of seeds and seed weight on the plant (r= 0, 61), number of seeds per plant and number of seeds in the bean (r=74), seed weight and beans on the plant (r=61), seed weight and beans one productive node (r=51).

3.3. Effect of year and place on seed productivity, selection of samples with high potential productivity.

In the 30s of the XX century, Engledow (pit. By Vavilov, 1966) tried to give the sowing area, the final result of all bio-physico-chemical processes occurring throughout the life of the plant in interaction with external conditions.

N.I. Vavilov (1966) attributed productivity to ecological traits (climatic, soil and seasonal conditions of the year) [14].

It is well known that the productivity of a variety or line is determined by its yield obtained under different climatic and agronomic conditions in most geographical locations.

One of the most important directions in bean breeding is the development of varieties with high and stable productivity. Data analysis shows that in order to obtain a stable yield, a system of varieties should be developed for each region, differing in morphological traits, maturity, and purpose of use [19;39;87].

In relation to productivity, many researchers note that this trait is the most variable [20;63;31]. Some authors believe that the least variable element of yield structure is the mass of 1000 seeds [21]. Others put the number of seeds in the pod in the first place in terms of ecological stability [79].

The yield is determined both by the biological characteristics of the test sample and the meteorological conditions of its cultivation. It depends mainly on the productive nodes, the number of seeds per plant, the weight of seeds per plant, the weight of 1000 seeds and finally on the degree of disease damage of the tested sample.

Our research has shown the rather large varietal differences in productivity of the studied assortment of varieties and forms of common bean, as well as their high variability by years in the study in the conditions of Apsheron. The amplitude of variability between them by years of the trial and on average for 5 years are as follows: 2010 г. -31-350; 2011 - 60-500; 2012. - 80-320; 2013 г. - 93-600; 2014 г. - 110-430; 2015 г. - 50-600.

The greatest influence on yield was the number of seeds in the plant. Not all yield components are directly related to seed weight. There is no direct effect of bean seed number on yield, but due to the significant indirect effect through plant seed number, the overall correlation with plant seed weight is quite significant.

Distribution of varieties and forms of common bean by grain yield under Apsheron conditions (2008-2012).

	Distribution of samples by phenotypic productivity classes						
	up to 100;	101-200;	201-300;	301-400;	401-500;	501-600;	601-700
2008	68	35	8	2	1	1	-
2009	59	37	15	3	1	-	-
2010	55	32	18	9	1	-	-
2011	22	36	32	16	5	2	2
2012	38	47	16	9	4	1	-
Totals: 115	42,1%	32,5 %	15,5%	6,8%	2,1 %	0,7 %	0,3%

More than 20 accessions exceeded the standard variety Yerli Piada in average yield, some of which are presented in Table 3.2. 3.2. The best varieties and lines.

Table 3.3.
Characterisation of sources of high seed production in beans, averaged over 2008-2013.

Variety	Number, pcs.					weight, g	
	productive nodes on the plant	of beans from the plant	beans one productive node	seeds from the plant	seeds in a bean	seeds from the plant	1000 seeds
к-13031	8	12	3	19	3	6,7	353
к-14534	12	16	4	36	4	10,5	294
к-15274	12	18	3	36	5	10,5	239
к-15275	14	6	2	11	3	3,6	334
т/15	6	16	3	32	6	11,67	371
т/16	6	14	3	14	3	7,18	513
т/17	8	16	4	36	5	19	528
т/18	6	18	4	34	4	18,19	535
AzePHA-7.5	8	9	3	22	3	8,93	406
AzePHA-20	12	10	3	35	5	16,55	415
AzePHA-14	8	9	2	28	4	11,11	397
azegri/34	8	11	3	28	4	9,43	337
Galibiyat	6	12	3	34	5	12,54	369
AzePHA-24	8	11	3	22	3	9,98	454

AzePHA-23	7	17	3	65	6	32,76	504
Т/27	10	30	5	150	7	25,65	171
т/37	7	7	3	14	3	9,33	450
Yerli Piyada	6	15	3	40	3	13,44	336
AzePHA-3	7	24	3	24	4	14,97	624
AzePHA-29	8	18	4	29	4	15,28	524

One of the main traits characterising the economic value of a variety is its yield, which depends on the number of fruit-bearing plants per unit area and seed weight per plant (productivity). In our experiments, most of the studied varieties had a fairly high plant survival rate during the growing season (2008 - 2013). Seed yield depends mainly on plant productivity. The latter is determined by several of its components - the number of productive node, seeds in the bean or bean filling and the weight of 1000 seeds (coarseness).

Fig.H. shows the interaction of form, variety 3 year (2010-2012) in the most productive 8 bean lines isolated under Absheron conditions. Of all tested varieties and form under Apsheron conditions, these 8 lines are the most valuable. Relatively high stability stands out especially 3 of ННХ-AFGO-27, t/37, AzePHA-20.

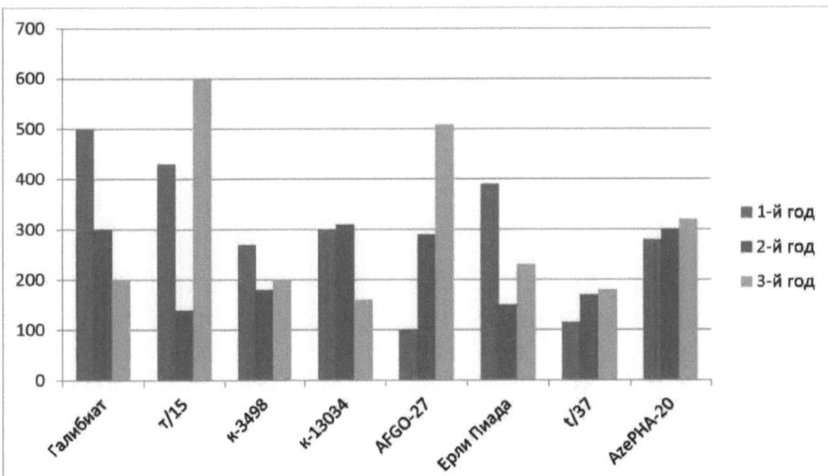

Figure 3: Effect of year conditions on grain yield variability of selected bean lines

3.4. Drought and salt tolerance of the source material .

Actual directions of seed including, high quality of beans in varieties of vegetable use. One of the important directions of breeding work is the creation of drought-resistant varieties. When breeding for drought tolerance, it is necessary to take into account the types of drought and the time of its greatest harm [40;24].

According to the degree of drought tolerance beans give sharp differences and by morphological, anatomical and biological features can be grouped into 3 groups of forms: xerophilic, hygrophilic and intermediate.

Beans occupy the fourth place among grain legumes after lentils, chinens and chickpeas in terms of resistance to drought. According to the data of P.F.Lvova for Kamennaya Steppe of Voronezh region, the transpiration height is not related to the degree of xerophilicity of forms: weak degree of stomata opening is characteristic for all ecotypes of common bean. The bean suffers from drought weaker in the sprouting stage; the critical period for it is the stage of grain filling.

According to G.M.Psarev, for Armavir, the resistance of common bean (variety "Robust") and tepary decreases with plant development: so, in the period before flowering plants withstand 10-17 days of drought, in the flowering period 5 days of drought are already fatal for them [35].

N.I. Vavilov (1966) emphasised that drought tolerance cannot be a specific trait characterised by constancy in all conditions [14]. The determining factors of plant reaction to moisture deficit are the development phase and previous growing conditions [82;31].

The relative drought tolerance of early maturing varieties is often reported in the literature [89;92;40].

Under emerging climatic conditions, with climate deviation towards warming and increasing as recurrence in drought tolerant varieties [3;83;41].

Drought tolerance is a complex trait associated with a number of physiological traits.

The yield of cultivated plants is largely determined by their adaptability to abiotic and biotic environmental conditions. Increasing the adaptability of cultivated varieties to specific growing conditions is the most important task of breeding. In breeding for adaptability it is supposed to identify the main limiting factors of the environment, regularly manifested in a particular growing zone for 4-5 years.

In Azerbaijan Republic 2010-2015 years are dry, and soil droughts are accompanied by high air temperature. During the years of our research, low precipitation was observed against the background of high temperatures. The year 2014-2015 was especially dry, which allowed us to evaluate the source material in specific weather conditions of Apsheron.

It is known that unfavourable environmental conditions, including drought and salinity significantly reduce all indicators of yield structure. Therefore, obtaining stable yields is associated with increasing the resistance of plant organisms. Drought and salinity are frequently occurring unfavourable environmental factors in some regions of Azerbaijan. At the same time, abiotic factors affecting crop yields such as low temperature during the germination period, as well as drought and high temperature during flowering and bobo formation occur in some regions.

In order to study and identify genetic sources of high stress-resistance, the reaction of bean samples to stress action by physiological parameter was evaluated. Experimental data show that the most drought-resistant samples were AzePHA-20, AzePHA-14, azrgri/68, AFGO-27. The degree of chlorophyll depression in these samples is completely absent. Samples t/6, azegri/68, k-3498, AzePHA-14 are identified as salt tolerant genotypes, in which the changes in chlorophyll amount under salt stress range from 112.4 to 146.1%. Genetic conditioned, inherited by plants level of resistance is the potential ability of varieties or forms to adapt to extreme environmental conditions.

Agricultural production needs adaptive bean varieties, capable to form a crop in any conditions, preserving the economic attractiveness of cultivation of the crop. It is necessary to search for new, drought-resistant bean varieties for the conditions of the Azerbaijan Republic.

The most resistant to drought conditions, taking into account the regression analysis of yield in dry years, were the following samples: Yerli Piada, Galibiat, AzePHA-20, t/6.

3.5. Chemical compositions of beans.

Chemical composition, nutritional and medicinal properties. Protein and its deficiency is a problem of national importance nowadays. Protein is the most important component of human food. Its deficiency causes physiological, functional disorders of the body: delay in growth and development, rapid physical and especially mental fatigue. Therefore, the level of people's welfare in a country is determined by the amount of protein consumed per capita per day. According to FAO, the norm of its consumption is 12% of the total caloric content of the daily human diet, or 90-100 g, including 60-70% of protein of animal origin. However, protein deficiency can be covered by vegetable protein at the expense of leguminous crops.

In our time, the low production of high-protein food of animal origin, their high cost motivates to increase the area under leguminous crops [67]. The most valuable food crop in this group of plants is bean [94]. The main supplier of plant protein - if counted on a global scale - is beans. About a dozen of them have been cultivated since ancient times. The age of remains of cultivated bean seeds found during excavations in Mexico is estimated to be V millennium BC. At the time of Columbus, Indians cultivated no less than 30 genera, one common bean. Food Beans are a staple diet and an essential source of protein for hundreds of millions of people, mainly in Latin America and Africa. Beans contain up to 25% protein, which is superior in nutritional value to many meats. In addition, bean protein is 70-80% digestible. Beans contain potassium, calcium, sulfur, magnesium, phosphorus, iron, vitamins Bi, B_2 _ B.< B_{1Y} C, E, K and PP, as well as essential amino acids. The seeds are rich in legumin and starch, the value of bean seeds is determined by the high protein content and its good digestibility by the human body. The protein digestibility coefficient is 86, higher than that of peas and lentils. The minimum protein content (15%) was found in the seeds of sharp-leaved bean (tepary), the maximum (33%) - in common bean under the conditions of cultivation on fertile soils of the Caucasus foothills.

The value of grain legumes is determined primarily by the high content of highly digestible protein in the shifts and other organs. The proteins of legumes include all amino acids necessary for nutrition - lysine, tryptophan, methionine, valine and others. High nutritional value of legumes is also due to the presence of a significant number of free amino acids, which are not part of the protein and therefore easily digested by the body. Such amino acids, including all essential amino acids, make up on average 4-5% of the grain weight.

The group of carbohydrates in vegetative and generative organs of bean is represented by starch, sugars - sucrose, glucose and reducing sugars. Polysaccharide stachyose is contained only in semi-ripe seeds, reaching up to 2 per cent by the time of maturity and rapidly consumed during seed germination, so it cannot be detected in the germinating plant. The content of sucrose and glucose also increases as the seed matures, reaching 3-4% for the former and 5-7% for the latter in mature seeds.

In particular, legumes such as chickpeas, beans and lentils are recognised foodstuffs in Azerbaijan. Numerous national dishes are prepared from the seeds of these crops. They serve as valuable raw materials in improving the quality and flavour of national bread products. Beans are a valuable food crop. Its seeds are used to prepare soups, pates, sauces, pilafs and many kinds of cold snacks. A tasty and nutritious product is quick-frozen green beans harvested at the stage of milk ripeness of the seeds. It contains complete proteins, sugar, mineral salts, vitamins A, Bi and C.

Nevertheless, these valuable food crops occupy insignificant areas, although there are great opportunities for their expansion. For example, in the Republic of Azerbaijan the area under bean crops is only 2.8 thousand hectares.

N.N.Ivanov (1927), and later Vogau (1931) and M.M.Kurgatnikov (1939) noted that most grain legumes, including beans, are characterised by relative stability of the chemical composition of seeds, in particular their protein content. N.N. Ivanov explained this feature by the fact that symbiosis with nodule bacteria provides favourable conditions for nitrogen nutrition of plants and causes their lower dependence on the content of mineral nitrogen in the soil [23]. Experimental data confirm that with the help of improved agrotechnics (by increasing nitrogen doses especially) it is possible to increase protein in grain by 1-5%. With the cultivation of beans under intensive technologies, along with the increase in yield, the quality of seeds is also improved. However, its content in seven beans is a hereditary trait. The negative correlation between yield level and protein content in grain is also well known. However, as Z. Barabash (1983) believes, from the practical point of view, the protein content in grain is not a sufficient indicator, the indicator of protein or lysine yield per unit area is more important. This indicator can be high enough in case of high grain yield per unit area.

At the same time, studies by I.A.Krasilnikov (1945), M.V.Fedorov and V.P.Podpolskaya (1951), M.V.Fedorov and S.V.Egorova (1957), L.M.Dorosinskii (1962) and other authors show that symbiosis of nodule bacteria with plants and the process of nitrogen fixation itself cannot be considered in isolation from soil-climatic conditions of legume plants cultivation.

Under the influence of fertilisers, the content of total and protein nitrogen increased, mainly due to globulin. The ratio between individual protein fractions was more strongly influenced by meteorological conditions than by fertilisers. M.I.Smirnova-Ikonnikova and co-workers (1962) established the dependence of protein fractional composition on geographical factors. The electrophoretic spectrum of albumin, globulin and glutelin protein fractions is specific for a variety and does not depend on growing conditions [85].

In addition, the factors under which yield and seed quality of these crops may be improved are also insufficiently investigated.

In this connection, the task of our research was a complex study of protein-rich leguminous plants beans - which sources play a significant role in the diet of the population of Azerbaijan.

On average, bean seeds contain one and a half times more protein than wheat, but less than seeds of peas, lentils and other legumes. Beans react sharply to growing conditions, which causes changes in the chemical composition of seeds. Studies by M.I. Smirnova-Ikonnikova and N.F. Pokrovskaya have shown considerable variability in protein content of bean species and varieties depending on growing conditions in different regions of the USSR. The experiments of M.I. Smirnova-Ikonnikova showed a significant increase in the lysine content of beans (variety Shchedraya) when phosphorus and nitrogen mineral fertilisers were applied [42].

Most of the nitrogen-free extractive matter of bean seeds consists of starch.

Studies by I.K. Murri (VIR) have shown [62] that bean beans contain significant amounts of provitamin A (carotene) and ascorbic acid. Ascorbic acid in bean beans accumulates as the beans mature: in the initial phase, 23-28 mg (per 100 g of raw bean weight) of ascorbic acid was determined. In bean seeds as they mature, carotene content drops sharply, while ascorbic acid content increases and reaches an average of 50 mg (per 100 g of raw weight) in the phase of technical maturity.

Bean seeds and immature beans contain trace amounts of the vitamin aneurin. Aneurin is present in both immature and fully mature dried bean seeds.

The following vitamins were also found in immature beans and mature seeds of common and lima beans: niacin (1.4 mg per 100 g in beans, up to 2.5 in seeds) and riboflavin (0.11 in beans and 0.22 in seeds).

Green beans and unripe bean seeds are rich in nutrients and physiologically active substances.

In 1950-1952, Prof. V.A.Chesnokov conducted a study of bean leaves and showed the possibility of extracting citric acid from them [93].

Mature bean seed ash contains the most potassium and phosphorus (almost three quarters of the total ash composition), followed by magnesium, calcium and sulphur. Sodium, iron and manganese are extremely low. Traces of elements such as copper, boron and molybdenum are found.

In our experiments, the sum of protein content of bean samples studied by us varied from 20.87 to 28.75%. The study of a large variety of common bean allowed us to identify 32 accessions with high, on average for the years of testing, protein content (23.12 - 25.06%, the standard has 22.68%) and with the lowest variability (up to 1%) of this indicator: k-13040; k-34; k-40; k- 35; k-14534, etc. In addition, there were 16 samples exceeding the standard in average protein content up to 6.07%.

Data on the content and distribution of protein in the grain of the studied varieties and form of common bean are given in Table 3.4.1.

Distribution of common bean accessions of local collection by grain protein content when grown under Apsheron conditions

Table 3.4.1.

	Protein content of grains by class (%)			
	17,31 - 23,0	23,12 - 25, 06	25, 12 - 27,06	27,50 - 28,75
	bush			
	26	18	8	2
	semi-hung			
	21	17	6	2
Bottom line: 100	47,0%	35,0%	14,0%	4,0%

The highest amount of protein in total dry matter was observed in sample k-36 - 28.75%; k-14361% - 27.18%; AzePHA-45 - 27.50%; AzePHA-18 - 27.06% *and others.* The least amount of protein was observed in sample azeqri/34, 17.31%; AzePHA-211-t, 21.00%; t/18, 21.37%; k-15275, 21.43% and others.

All studied bean species and varieties are characterised by a full set of amino acids, but differ significantly in their content. However, the nutritional value of food products is determined not only by the amount of protein included in their composition, but also by its completeness, i.e. amino acid composition.

Legumes serve as a good source of a number of essential amino acids to improve the nutritional adequacy of diets. The use of legumes to increase the protein content of bread and to supplement a number of missing essential amino acids in other foods is currently being promoted [8].

In this connection, the study of amino acid composition of proteins of local varieties of legume crops is of great interest.

The amino acid composition of food legumes of Azerbaijan determines their biological value and the efficiency of their use as one of the important elements of improving the protein balance of the country [77;78].

The selected samples can be used in breeding work for grain quality (Table 3.4.2)

Table 3.4.2.

Results of biochemical analysis of common bean samples

Naz. Obraztsov	In No-air-dry weight			mg in 100 g	
	total nitrogen	Protein	Ash	tryptophan	Lysine
к-36	4,6	28,75	4,25	240	1030
AzePHA-45.	4,4	27,5	4,56	245	1128
AzePHA-18	4,33	27,06	4,49	250	947
AzePHA-20	4,09	25,56	4,02	196	839
azeqri/68	3,91	24,43	3,45	200	1178
к-14361	4,35	27,18	4,41	235	1128
к-15274	3,91	24,43	3,79	210	871
AzePHA-15	4,01	25,06	4,3	230	982
к-40	3,98	24,85	3,75	175	849
st. Galibiat	3,80	23,75	4,14	185	974
Yerli Piada	3,63	22,68	3,90	210	928
AzePHA-3	4.20	26,18	4,30	230	982
т/7	4,30	26,87	4,10	250	947

3.6. Diseases and Pests of Beans.

Grain legumes are damaged by pests both specific, feeding only on legume crops, and multivorous, damaging not only legumes, but also other agricultural crops. Production of grain legumes is significantly damaged by diseases - fungal, bacterial and viral diseases

Diseases drastically reduce the bean crop and sometimes completely destroy it. Depending on what pathogens cause bean diseases, they can be divided into 3 groups: 1) fungal diseases (anthracnose, fusariosis, etc.) caused by fungal parasites; 2)- bacterial diseases (bacterioses) caused by various types of bacteria; 3) viral diseases (mosaic, etc.) caused by viruses. All local and breeding bean varieties are affected by diseases to some extent. The resistance of some bean varieties to diseases is relative.

Fungal diseases - anthracnose. The causative agent is *Colletotrichum Lindemuthianum* Br. Et Gav. Anthracnose is the most damaging fungal disease for beans. The disease particularly affects beans, on which rounded dark grey or brownish, slightly depressed spots with a dark border and a dark red stripe appear.

Often, holes are formed on the beans in place of the spots, through which the disease affects the seeds. Sprouts and adult plants are also affected by anthracnose. Anthracnose affects vegetable bean varieties most severely in the non-Black Earth zone, especially in the closed ground.

The disease is transmitted through soil and seeds. In bean seeds, the mycelium remains viable for 5-7 years. The main source of the disease is the undecomposed remains of stems and leaves from last year's harvest, which retain the mycelium of the fungus. If you sow diseased anthracnose bean seeds in the soil, then on the seedlings you can immediately notice the manifestation of this disease (on the seedheads), then the disease passes to the leaves (veins and petioles are affected), and later - on young beans (green spatula). The spores of the fungus begin to germinate at 10°, but the optimum temperature is 2°.

Old plant tissues are more resistant to anthracnose than young ones, the fungus damages only a few cells and dies off. The fungus overwinters in the seed coat or in the seed pods. The optimum acidity for development is pH 5.5-6, which is the normal acidity of bean leaf and young stem juice.

Dry root rot, or fusarium. The pathogen is *Fusarium* Maptii App. Et Wr. var *phaseoli* Burk. The disease is detected soon after the emergence of bean seedlings. External signs of diseased seedlings are the appearance of an overstretch on the sub-seedling above the root neck. The base of the stem becomes thinner than the upper part (above the tug). The fungus then spreads from the root neck up the stem and infects the vascular system, which disrupts the water supply; the plants wilt, and the leaves turn brown and fall off. As a result, plants die or drastically reduce productivity. In dry weather, plants die from this disease (according to the observations of P.M. Sheterenberg in Odessa region), much faster than in wet weather [42].

The main source of root rot infection in beans is soil, but the disease can also be transmitted through seed.

Ascochytosis. The pathogen is *A^occhyla phaseolorum* Sacc. It mainly affects common bean, but also occurs on golden and multiflowered beans. The disease affects stems, leaf petioles, beans and seeds. Elongated brown spots form on stems and leaf petioles, and oval, transparent, oily, light brown spots form on beans. Seeds affected by ascochytosis are puny, with blurred spots of ash colour. This disease is not widespread in our country.

Cercosporosis. The pathogen is *Cercospora cruent* Sacc. It affects leaves of golden and angular beans. External signs of the disease are appearance of brownish-brown spots on leaflets with browning of leaves. The disease was recorded in the Azerbaijan SSR and Central Asia [42].

Bacterial diseases (bacterioses). Pathogens -*Xanthomonas phasoli* (E. Smith.). Dowson, Corynebacterium flaccumfaciens (Hedges) Dowson and others. Bacteria causing the disease of beans are very mobile, resistant to drying, but do not tolerate high temperatures and die at 50 °, belong to the aerobes.

Bacterioses affect leaves, stems, also beans and seeds. They most severely affect beans in the main zone of their distribution, especially in the steppe areas of Ukraine and Kuban. More than 10 bacterial diseases of beans have been described, but bacterial blotch, bacterial wilt and bacterial halo are the most severe. Bacteriosis affects young beans, leaves and stems with a brown spot; as they mature, the spots dry up and become brownish-brown; the spots on beans are usually larger than those caused by anthracnose. It was found that varieties and forms of common bean are affected by five pathogens of bacteriosis, multiflowered bean - by two [42].

Viral bean diseases seem to be widespread. Characteristic signs of these diseases are: twisting of leaves, deformation of leaf lamina (swellings), leaf curl, shortening of internodes, discolouration of leaves (alternation of dark green lamina areas with yellow), dwarfism of plants, etc. The most common viral diseases are mosaics: curly, stripe and vesicular. Other mosaics are also known - green, yellow.

Resistance to common mosaic is a recessive trait depending on the combination of 2 genes. Common and yellow mosaic viruses affect a number of species of the American and Asian groups. The wrinkle mosaic virus is extremely highly specialised: it was found by Nelson on only one variety of common bean ('Stringless Refugee'). Resistance to common mosaic is inherited independently of resistance to yellow mosaic. Thus, the cultivars listed above that are resistant to common mosaic are at the same time susceptible to yellow mosaic.

Important measures to control viral diseases of beans are: development of resistant varieties, sowing with seeds obtained only from healthy plants, removal (before flowering) and burning of diseased plants and control of insects - vectors of viral diseases.

The main in the fight against diseases of leguminous crops is agrotechnical and organisational measures. The main ones include: observance of crop rotation, placement of leguminous crops at a permissible distance from perennial grasses. Grain legumes should not be returned to the field occupied by legumes earlier than in 3-4 years.

Seeds for winter storage should be put into conditioned storage, especially in terms of purity and moisture. Seed humidity should not exceed 14%, seeds should be free of any impurities and stubby grain.

Bean pests. Bean bean weevil *(Acanthoscelides obtectus* Say). The most dangerous bean pest, damaging seeds both in the field and in granaries. The damage is caused by beetle larvae developing from eggs laid by females either directly on seeds or bags (in granaries) or on mature beans (in the field). Larvae penetrate inside the seed, and several 1- 4 (up to 28) larvae may develop in one seed. The larval development period averages 24 days. The total number of bean weevil generations is at least 4, with 2 of them developing in the field and the rest in autumn and winter in granaries. The food specialisation of the bean weevil was studied by F.K. Lukyanovich. He found that the bean weevil can develop on five cultivated bean species [42].

Gall (root) **nematode** *(Meloidogyne marioni* Cornu). Damages roots of all bean species, especially golden beans. The damage is caused by larvae; they penetrate into the cortical part of the root and cause abnormal growth of its cells (cortical parenchyma), resulting in abnormal tissue development on roots with formation of swellings - galls.

The nematode is widespread in some areas of the Ukrainian, Georgian and Azerbaijani SSRs, as well as in the Central Asian republics.

Our crops were mainly affected by bacterial halo.

The degree of damage caused by this disease depends mainly on weather conditions. If conditions are favourable for the disease, the damage is much more severe during flowering and fruiting.

Under the conditions of the zone under consideration, in some years there is a slight spread of anthracnose caused by the fungus *Colletotrichum Lindemuthirinum* Br et Gav. For example, 'Galibiyat', 'YerliPiada', 't/15', etc. were weakly affected by anthracnose.

During the years of research we isolated 14 samples of grain and vegetable beans resistant to fungal and bacterial diseases: k13034, AzePHA-20,

Resistance of bean samples, depending on the species, was expressed to a different degree in the years of research. The analysis showed that common grain bean was more resistant to bacteriosis.

Pest damage was observed only by bean grain beetle - *Acanthoscelides obtectus* Say. Saxa fibreless 615", "k-15274", "k-13036" were more severely affected by pests.

CHAPTER 4

4. Parameters of the experimental model of bean variety

Of a number of directions in the breeding of common bean, the first place belongs to selection for immunity. Breeding is carried out mainly in the direction of developing varieties highly resistant to fungal, bacterial and viral diseases by methods of individual selection and hybridisation. Another direction of breeding is breeding for early maturing, drought-resistant and cold-resistant varieties. The third direction is selection for quality.

Parameters of the experimental model of bean variety adapted to the conditions of the Republic of Azerbaijan. Taking into account the developed and scientifically substantiated optimal parameters for the main quantitative traits, the models of varieties for early and middle ripeness groups (bush and curly) were proposed.

4.1. Relationship between productivity and other economically valuable traits.

Knowledge of the regularities of the influence of each trait on productivity, as well as their variability in a particular zone of cultivation, allows to identify the value of each trait on productivity, makes it possible to evaluate genotypes and make changes in the selection practice and in the parameters of structure elements and morphobiological traits when modelling new varieties. To find out the most effective selection and evaluation of bean breeding material, correlations between plant productivity and elements of yield structure are important.

Biometric analysis of plants of the collection showed that the variability of the main traits determining productivity was due to genotypic diversity and the influence of year conditions. The average values of the coefficient of variation of the traits "number of seeds, beans and productive nodes per plant" and "seed weight per plant" indicate a strong degree of their variability. The highest level of variability was observed for the trait "seed weight per plant" with the maximum value. The traits "number of seeds in a pod, weight of 1000 seeds" were characterised by an average level of variability.

The magnitude of correlation coefficients obtained in the studies of different authors often seems to be related both to the set of studied varieties of samples and to the growing conditions.

The number of seeds in the pod has a positive correlation with the number of seeds per plant between 1000 seed weight and the number of seeds per plant, 1000 seed weight and the number of seeds in the pod showed a weak negative correlation, and a very weak both negative and positive relationship between seed weight and stem length, 1000 seed weight and the number of seeds in the pod and the number of beans per plant.

Thus, in our experiments, in all years of research, a reliable relationship at the inter-variety level was found between seed weight per plant and stem length, seed weight per plant and number of beans per plant, seed weight per plant and number of seeds per plant. To a lesser extent, productivity was related to the number of seeds in the bean (Figures 4.1. .4.8).

The development of variety modelling principles is based on knowledge of the range of variability of quantitative traits in this particular zone of cultivation of possible correlations of the most important traits in the formation of yield.

The range of variation makes it possible to establish the limits of manifestation of bean traits in the studied conditions, and the found correlations between them show what traits should be selected for.

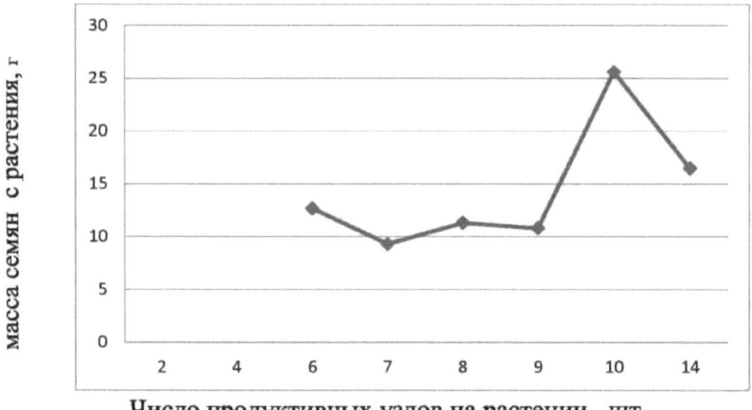

Figure 4.1. Relationship of productivity of one plant with the number of nodes on the plant

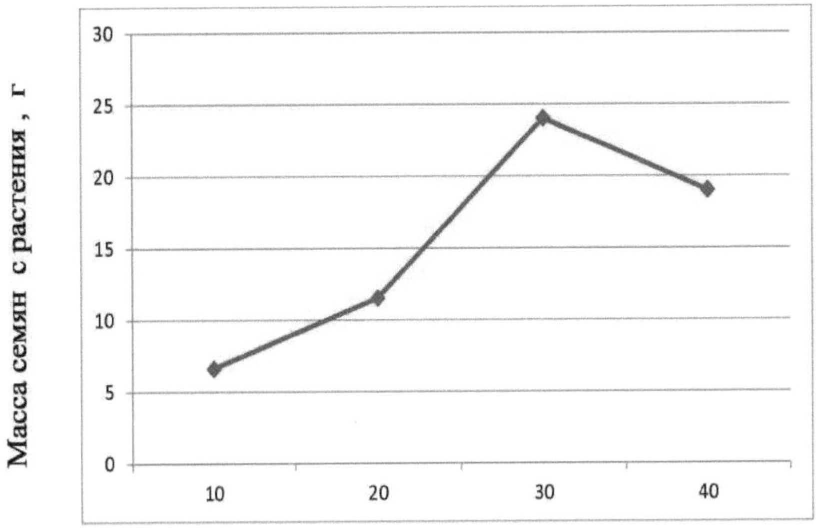

Figure 4.2. Relationship between productivity of one plant and number of beans per plant, pcs.

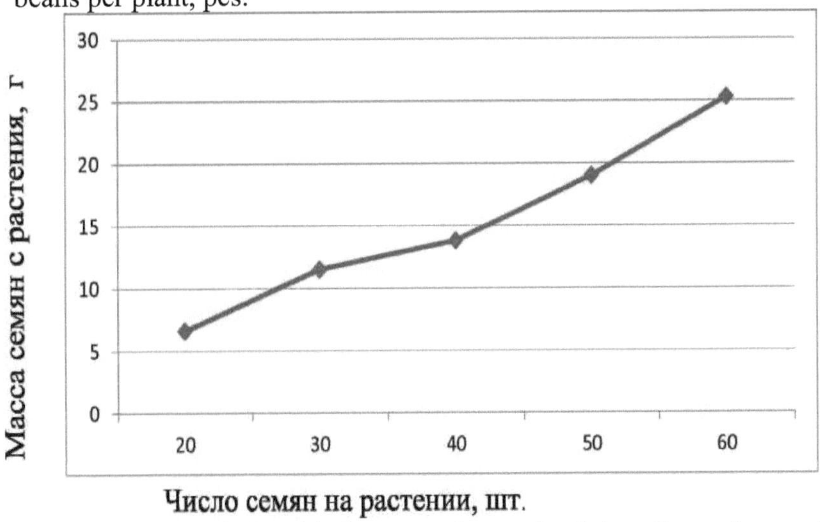

Figure 4.3. Relationship between the productivity of one plant and the number of seeds on the plant.

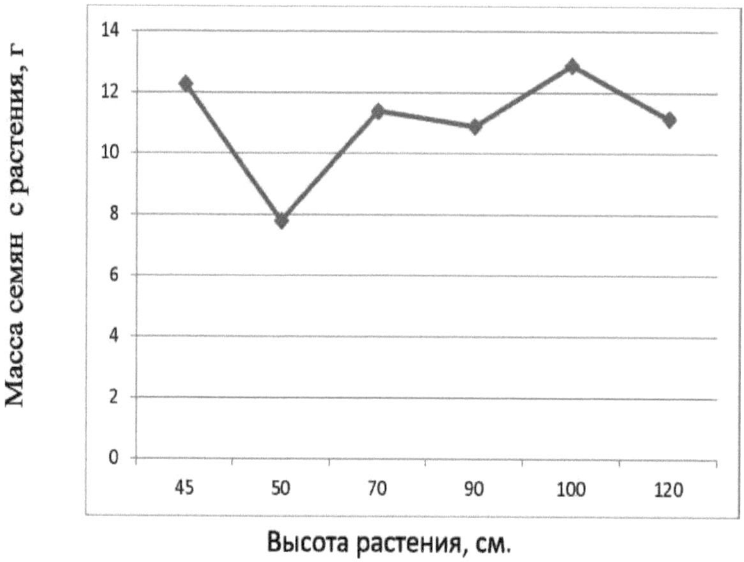

Figure 4.4. Relationship of productivity of one plant with plant height.

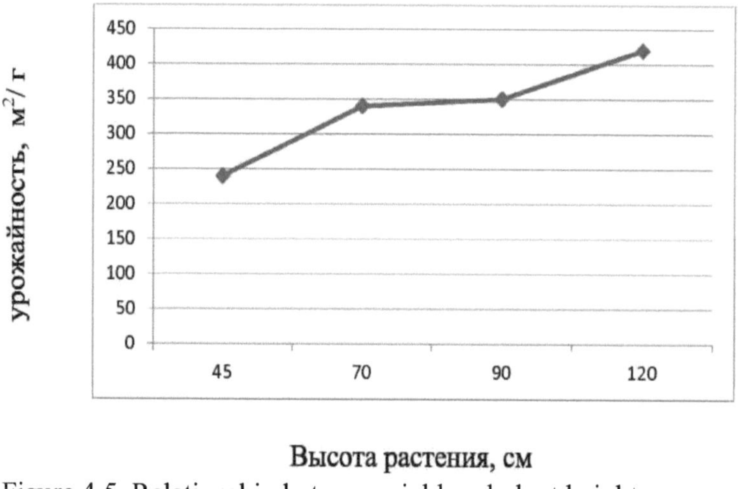

Figure 4.5. Relationship between yield and plant height

Figure 4.6. Relationship between yield and number of productive nodes.

Figure 4.7. Relationship between yield and number of beans per plant.

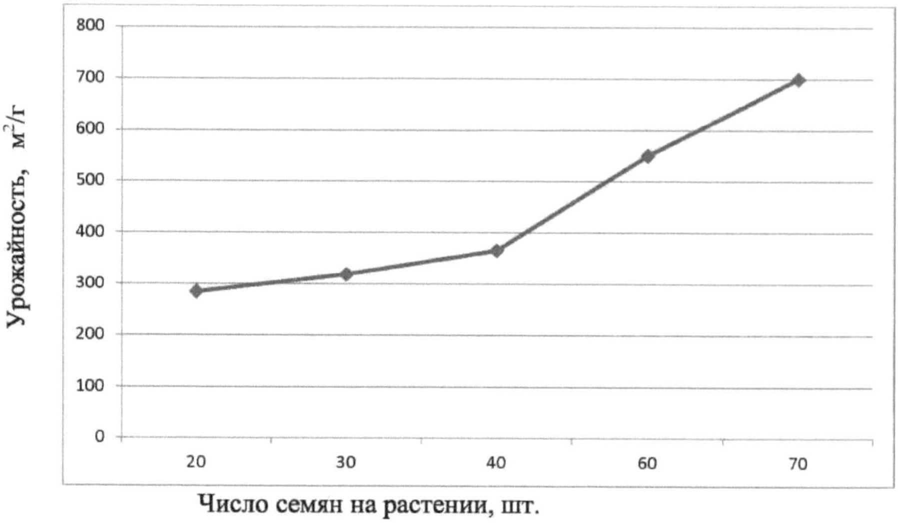

Figure 4.8. Relationships yield number of seeds per plant.

Graphical analysis of the interaction between yield and plant height showed an increase in yield per square metre, (Figure 4.5.) as well as productivity per plant (Figure 4.4.).

In Figures 4.1 and 4.6, a direct relationship between the number of productive nodes and productivity per plant and yield per plot can be seen. For the parameters of the model variety of early ripeness group, the number of productive nodes should be within (16-18 pcs.). The optimal number of beans is (16-24) (Fig. 4.2). The number of seeds per plant also significantly affects the yield per m^2 , as well as the productivity per plant (Fig. 4.8;4.3).

It has been established that it is possible to select for high productivity by one plant, as there is a correlation between productivity of one plant and yield from one m -a.^{2}

Seed production is a ratio of many components. In the end, it is influenced by the number of seeds per plant and the weight of 1000 seeds.

Seed productivity is divided into two categories: potential and actual (actual) seed productivity [17]. Potential seed productivity is the maximum possible number of seeds that a plant, population, or phytocenosis can produce in a certain period of time, provided that all seed nuclei laid in flowers can form mature seeds. Actual seed production
productivity is the number of normally developed seeds per accounting unit [60].

In our experiments, on average for the years 2011-2014, data analysis shows that on average for 4 years, the varieties with large seeds had the highest productivity, while the small-seeded varieties had significantly lower productivity.

Consequently, the combination of elements of yield structure in medium-seeded and large-seeded varieties in the conditions of Azerbaijan is the most favourable. Thus, T/27 has the largest number of seeds on the plant, but due to the low weight of 1000 seeds on productivity takes one of the last places.

4.2. Parameters of experimental model of bean variety adapted to the conditions of Azerbaijan Republic.

The development of variety modelling principles is based on knowledge of the range of variability of quantitative traits in a particular cultivation zone, possible correlations of the most important traits with yield, and the contribution of individual traits to yield formation.

The range of variation makes it possible to establish the limits of manifestation of bean traits in the studied conditions, and the correlations found between them show which traits should be selected for.

Taking into account the developed and scientifically substantiated optimal parameters for the main quantitative traits, variety models for early and middle ripeness groups were proposed (Tab. 4.2.).

Table 4.2.

Indicator	The Yerli variety is a regionalised variety Pia yes.	Model grade AzePHA-20	Regionalised variety Galibiat	Model grade AzePHA-3
seed yield m/y	390	280	400	600
branch count	3	3	3	3
number of productive nodes per plant	6	12	6	7
number of beans per plant	15	10	12	24
number of seeds in the pod	3	5	5	4
number of seeds per plant	40	35	34	64
1000 seeds weight, g	336	415	294	624
plant height, cm	30-35	50-70	80-90	80-90
height of lower bean attachment, cm	11	16,5	14,6	20
yield quality: protein in seeds,%	22,68	25,06	23,75	26,18
drought resistance	average	high	average	high
pest damage	average	average	Medium	Medium

For the model of medium group (bush) bean ripeness, plant height parameters of 50 to 70 and lower bean attachment of 15-28 cm were determined.

CHAPTER 5

CONCLUSION

The problem of mastering and expanding plant genetic resources in breeding has long existed. Its relevance is steadily increasing due to the identification of breeding and the constant need to improve and enrich the varietal gene pool. This was foreseen by N.I. Vavilov: he was one of the first who took measures to mobilise the conservation and comprehensive study of the existing varietal and species diversity of agricultural plants.

The need for systematisation of the gene pool of breeding has especially increased in the last two decades due to the work on creation of state *(national)* and international gene banks of agricultural plants. The main principle in the development of the system is the grouping of plant species and forms according to the possibility of their use as source material in breeding programmes for specific crops.

Bean is the main leguminous crop in our country widely cultivated in different soil and climatic conditions.

The new strategy for the development of crop seed breeding in the Republic of Azerbaijan for the period up to 2018 directs breeding to the creation of high-yielding varieties, stable forming high quality yields, capable of resisting environmental stress factors and efficient use of anthropogenic and natural resources, to improve existing and develop new technologies of breeding process with the involvement of modern research methods, bioengineering, effective donors and sources of valuable traits.

Cultivated varieties of grain and vegetable beans are insufficiently technological. At present, a number of varieties with the property of non-shattering seeds, resistant to..., but the successful introduction of such varieties into production is hindered by their relatively low yields.

Modern breeding should have mobility in solving actual practical problems, for which it is necessary to have a variety of source material, which can be the source of both individual traits and their various combinations.

This work is devoted to a comprehensive study of bean collection in the conditions of Absheron in the Institute of Genetic Resources of ANAS

Azerbaijan 2007-2015, 130 bean samples were studied.

Peculiarities of variability and interrelations of productivity traits of different bean morphotypes were determined. The sources of main economically valuable traits were identified from the studied varieties belonging to different bean morphotypes. Early maturing, high-yielding, drought-resistant varieties were identified. As a result of development of different populations the initial material with a complex of economically valuable traits and properties was created. This selection material is of great interest for breeding and agricultural production.

In 2005-2015, 130 bean varieties and bean accessions were described for qualitative traits. Description of bean varieties and samples on morphological traits was carried out in several stages: during sprouting, flowering, technical and biological ripeness. Sources of early-sprouting and medium-sprouting bush forms were selected.

Correlations between morpho-biological traits of varietal samples of the collection have been established, which can be used in breeding work when creating new bean varieties.

Bean cultivation is characterised by a very complex technology, differentiated according to many factors. It is crucial to take into account the actual and expected weather conditions.

The development of variety modelling principles is based on knowledge of the range of variability of quantitative traits in a particular cultivation zone, possible correlations of the most important traits with yield, and the contribution of individual traits to yield formation.

The range of variation makes it possible to establish the limits of manifestation of bean traits in the studied conditions, and the correlations found between them show which traits should be selected for.

LITERATURE

1. Abdullaev A.A. Study of variability and correlative relations quantitative traits in cotton // Uzbek Biological Journal.1975.No.5.
2. Adamova O.P. Influence of growing conditions of grain legumes on seed formation. - Proceedings on applied botany, genetics and selection. 1971, vol.45. vol.Z.
3. Alabushev A.V. Adaptive potential of varieties of grain crops. // Grain legumes and cereal crops,- Orel: VNIIZBK, 2013. № 2.-C 4751.

4. Amirov L.A. Agrobiological study of beans of different origin in conditions of Sheki - Zakatala zone of Azerbaijan SSR. Author's thesis for the degree of Candidate of Sciences. of agricultural sciences. Leningrad, 1981.
5. Asadova A.I. State of breeding work with vegetable and grain beans in Azerbaijan. // New and non-traditional plants and prospects for their use. Proceedings of the XI International Symposium. Pushchino, 2015. C.285-287.
6. Asadova A.I., Amirov L.A. Study of economically valuable traits of samples of different bean species in Azerbaijan. Introduction, conservation and use of biological diversity of cultivated plants. // Materials of XI international scientific and methodical conference. 4.2. Makhachkla, 2014. C. 98-101.
7. Badina G.V. Cultivation of leguminous crops and weather. Л.- 1974. C. 87108.
8. Bakhitov R.F. Study of the main stages of the process of formation of technological properties of chickpea and bean under the influence of various agrotechnical factors in the conditions of Azerbaijan SSR. autoref.dissertation, submitted for the degree of Candidate of Agricultural Sciences. Baku, 1969.
9. Balashova N.N., Zeman G.O. Vegetable growing.-Tashkent. UKITUVCHI, 1981 P.350-353.
10. Buravtseva T.V., Egorova G.P. Collection of beans VIR-100 years // "Grain legumes and cereals", № 4 - 2012. C.46-52
11. Bodnar G.V., Lavrinenko G.T. Leguminous crops. M.- "Kolos", 1977. C.12-14; 36-44; 134-139
12. Borevic S. Principles and methods of plant breeding (translation from Serbo-Croatian). M. "Kolos". - 1984. - 344 c
13. Vavilov N.I. Theoretical bases of selection. M., "Nauka", 1987, 511 p.
14. Vavilov N.I. Genetics and breeding. // Selected Works. Kolos 1966. T.V. C. 451-465
15. Vavilov P.P. Plant growing. / edited by P.P.Vavilov - M.: Agropromizdat, 1986.- 512 p.
16. Vavilov N.I. Izb.trudy. T.Sh.M.-L., "Nauka", 1962, -S.42-51.
17. Vainagii V.I. On the methodology of studying the seed productivity of plants. //Botanical Journal. 1974. VOL.59.NO.6.P.826-831.

18. Verbitsky N.M. Pea breeding in conditions of insufficient moisture. / N.M.Verbitsky // Bulletin of the Russian Agricultural Institute. - M 1994. № 2
19. Verbitsky N.M. Pea breeding in the conditions of the North Caucasus. - Rostov-on-Don, 1992. C. 259
20. Verbitsky N.M. Selection of pea for productivity // Scientific and Technical Bulletin of VASHNIL. Siberian Branch Novosibirsk, 1981. Vyp. 6- 7,- c.96-99
21. Verbitsky N.M. Source material for pea breeding in Rostov region. Author's thesis of candidate of s.- x. sciences. - L., 1969. C. 3-16
22. Vetrova E.G. Pea selection for high productivity. // Scientific, technical bulletin. All-Union. Selekts.-Gen. Institute.- Odessa, 1986. Vyp. 4.- c.38-41.
23. Vogau N.A. Experience in accounting of crop quality. "Sots. grain. khoz." № 6, 1934
24. Gamzikova O.I., Kalashnik N.A., Gudinova L.G. Some genetic aspects of drought resistance and soft spring wheat // Problems of breeding of agricultural plants. - Novosibirsk. 1983.- C. 90-103.
25. Georgiev D. Testing bean varieties for suitability for mechanised harvesting. 1987
26. Golban N.M., Rassokhina A.I. Methods and results of bean selection for suitability for mechanised harvesting. Selection and seed production of field crops in the MSSR. 1987. C. 47-54.
27. Gorina E.D., Anokhina G.A. Towards selection of buckwheat for early maturity. // Breeding and seed production.- M., 1980.- № 2.-S. 16-17.
28. Davletov F.A. Selection of non-sucking pea varieties in the conditions of the Southern Urals. - Ufa, 2008. C. 236.
29. Davletov F.A. Methods and results of pea breeding. Bashkortostan.- Ufa: LLC "Epokha". 2006, c.92.
30. Davletov F.A., Popov B.K. Selection of pea varieties for the conditions of the Southern Urals. // Grain farming. - M., 2004.- № 8 - P. 7-8.
31. Davletov F.A. Selection of grain pea in the conditions of the Republic of Bashkortostan. - Ufa, 1993. C. 3-51
32. Dekaprelevich L.L. Bean varieties cultivated in Georgia. Notes of the Scientific and Applied Department of the Tiflis Botanical Garden, v. IV, 1925

33. . Dekaprelevich L.L. and Menabde V.L. To the study of field crops of Western Georgia. 1. Racha. Notes of the Scientific and Applied Department of the Tiflis Botanical Garden, v. VI, 1929
34. Dekaprelevich L.L. Beans. M., "Kolos", 1965
35. Dietmer E.E. Cultural Flora of the USSR. Edited by Prof. E.V.Wolf. Leningrad, 1937. C.458-533.
36. Ditmer E.E. To the question about the origin of the cultivated bean. On Prikkl. Бот.- Л.,1931.-Т.23-Вып.5.-С.309.
37. Dorofeev V.F. Wheat of Transcaucasia. //Tr. Po prik. bot., gen. and sel. Л - 1977. Т.47. - Vol. 1. - С.5-202.
38. Zhuchenko A.A. Ecological and genetic problems of plant breeding. *And* Agricultural Biology. - №3. 1990, с. 3-23
39. Zadorin A.D., Yakovlev V.A. Results and prospects of pea breeding in Russia. // Breeding and seed production.- M., 1994.- № 1. - C.2-5.
40. Zelenov A.N. Physiological bases of selection of leguminous crops for drought resistance. // Resistance of leguminous and cereal crops to unfavourable environmental factors and ways to improve it. - Orel. 1982.- C. 4-16
41. Zotikov I.V. Scientific support of grain legumes and cereal crops production in the Russian Federation: state and prospects. // Grain legumes and cereal crops.- Orel : VNIIZBK, 2003.- № 2.- P. 10-17
42. Ivanov N.R. Beans. 1961
43. Ilyina L.G. Creation of spring wheat varieties with high grain quality. // Vestnik s.-kh. nauki.- 1984.№10.-S.101-106
44. Kalaidzhieva S. Influence of sowing date and method on yield and grain quality of soya bean. Rasten. Nauki. 1989. 26. 1. 15-20.
45. Kevkhshivili V.I. Features of technology of cultivated field crops in the rainfed areas of eastern Georgia (Gore-Kakheti plateau). Author's thesis for the degree of Doctor of Agricultural Sciences. Kharkov, 1987.
46. Kolompikova G.D., Vasyakin N.I., Kiprev Y.N. Timing, norms and methods of sowing in the steppe zone of the Omsk region. // Intensifakasiya cultivation of agricultural crops in Zapodnaya Siberia. 1988 C. 109-113.
47. Korenev G.V., Podgorny P.I., Scherbak S.N. Plant growing with the basics of selnction and seed production. - M.: Kolos, 1983.- P.212-216.
48. Krylov A.V. To the questions of creating vegetable bean varieties with high sustainable yields and high quality products. - "Notes of the Voronezh

Agricultural Institute", 1939, Vol. 17, No. 1.
49. Kuperman F.I. Morphophysiology of plants. - Moscow: Vysh. Shkola, 1973.P. 256
50. Lelli Y. Selection of wheat. - M. Nauka, - 1980.- 384 p.
51. Leshchenko A.K., Matushkin V.A. Soybean selection for early maturity and high productivity. Breeding and seed production. 1989.67, c.39-44.
52. Litun P.P., Zozulya A.L., Dragovtsev V.A. Solutions of selection problems on the basis of ecological and geographical model of quantitative traits. //Breeding and seed production: interved. themat.nauchnogo sb.- Kiev: Urozhay, 1986.- Vyp. 61. C. 3-13.
53. Likhochvor V.V., Petrichenko V.F., 1vashuk P.V., Korshichuk O.V. Roslinnittvo.- Lviv. Technologies of cultivation of agricultural crops. (120 crops)/- Lviv NVF "Ukraashsky Technologi", 2010- 1081 p.
54. Lukyanenko P.P. Methods and results of winter wheat breeding in Kuban. №. 1967
55. Makasheva R.Kh. Cultural Flora of the USSR. ed. by O.N.Korovinya - L.: Kolos. 1979.- Vol.4. Ch.1.- 321 p.
56. Makasheva R.Kh. Peas.- L.: Kolos, 1973.- 312 p.
57. Mammadov M.N. Study of bean samples of Azerbaijan and VIR collection in order to identify the best of them for selection. Author's thesis for the degree of candidate of agricultural sciences. Baku, 1968
58. Manafov S.A. Influence of fertilisers on pea and chickpea yields in the conditions of Azerbaijan SSR. Baku, 1968
59. Maksimovich M.M. Selection and seed production of field crops. M. I zd. s.-kh. literature, magazines and posters. -406.
60. Methodical instructions on seed science of introduced plants. N.V.Tsitsin. M.: Nauka, 1980.64 p.
61. Minyuk G.M. Beans./Minsk: Urozhay, 1991, - 93 p.
62. Murri I.K. Dynamics of vitamins in vegetable legume varieties. In the collection "Biochemistry of fruits and vegetables". № 2. Izd. of the Academy of Sciences of the USSR, 1951
63. Naumkina T.S. Variability of the main economically useful traits in pea. // Collection of scientific papers on applied botany, genetics and breeding. Botany, genetics and breeding. - M., 1988. C. 121-125
64. Nosavotsky A.I. Wheat. *8* Biology. 1965, c. 586.
65. Nettevich E.D. Increasing the productivity potential of grain crops -

precocity. // S.-H. Biology, No.1. - 1982.- C.9-13.
66. Ovcharuk O. Agroecological characteristics of varieties of kvassol and their productivity in the conditions of Zakhchychny Liyustep // 36 scientific works. УНУС.- Вип.84.Ч.1.-Умань,-2014.-107-112
67. Ovarchuk O.V. Characteristics of sorivs to Baconi in utovyh JlicocTeny zahshchnogo. B: 36. Nauk.pratsy 1nstitut buenergetichnykh kultur! tsurkovyh buryak! Krzw, vip. 17 (1), 2013. c. 236-239.
68. Obraztsov A.S. About some biological aspects of the problem of selection for early maturity. // S.-agricultural biology.- M, 1983.- №.- p.3-11.
69. Petrichenko V.F., Movran K.L Influence of the method of s1vbya and density of the plants on the starchy productivity of xBaconi plants of extraordinary V: Fodder and forage production: M1zhv1d. Thematic. nauk. 36. Vshnitsa Vin. 2010. 67, c.64-69.
70. Pivovarov V.F. Vegetables Rossi.- M., JSC "Russian Seeds", 1994 p.6871.
71. Polyanskaya L.I., Chekrygin P.M. Selection of beans for suitability for mechanised harvesting,// Selection and seed production.- K.: Urozhay, 1983.- Vyp.53.- P.35-37.
72. Polyanskaya L.N., Zaginaylo N.I. New bean varieties // Selection and Seed Production.- No.3, 1991.-S.39-40.
73. Pryanishnikov D.N. Izbor. op. cit. M., "Kolos", 1963
74. Pryanishnikov D.N., Yakushkin I.V. Plants of field culture (private farming) OGIZ. Selkhozgiz.-M., 1938. C. 285-287.
75. Rabotnov T.A. Biological observations on subalpine meadows of the North Caucasus. // Botan. Zhurnal. 1945. T. 30. № 4. C. 167-177.
76. Razumov V.I. Environment and developed plants. L.-M., Selkhozizdat, 1961.
77. Rafiev E.B., Asadova A.I. Study of protein and tryptophan content in collection samples of vigna and vetch. Introduction of non-traditional and rare plants. // Materials of the X International Scientific and Methodical Conference in memory of Academician RASHI Nemtsev Nikolai Sergeevich. Vol. 2. Ulyanovsk, 2012. C. 175-177.
78. Rafiev E.B., Gasimov G.G., Asadova A.I., Allahverdiev T.I. Study of the content of nitrogenous substances in seeds of collection bean samples. New and non-traditional plants and prospects for their use. // Proceedings of the IX International Symposium. VOL. III. Pushino, 2008.- P.235-238.

79. Rodin E.A. Correlation of traits in pea plants. // Proceedings of Kirov Agricultural Institute.- 1981.- P.170-175
80. Rodnikov N.P., Kuryukov I.A. Smirnov N.A. Vegetable growing. M. Kolos, Obraztsov A.S. About some biological aspects of the problem of selection for early maturity. // S.- x. biology. - M., 1983.- № 10.- P. 3-11.
81. Rudenko A.I. Determination of phases of development of agricultural plants. Moscow, Izd. of the Moscow Society of Nature Experimenters. 1950.
82. Rybnikova V.A. Yielding pea varieties under drought conditions. // Breeding and seed production.- M., 1981, № 5.- P.26-28.
83. Savchenko I.V. Innovative development of crop production in modern conditions. // Grain legumes and cereal crops. - Orel: VNIIZBK. 2013.- J62.-C.4-9.
84. Selection and seed production of field crops. L.: Kolos. - 1988.- 335 c.
85. Smirnova-Ikonnikova M.I. Characterisation of plant resources of grain legumes by quantitative and qualitative composition of protein. In the book: "Proteins in industry and agriculture". Izd. of the Academy of Sciences of the USSR, 1952
86. Stepanov V.N. Relation of agricultural plants of field culture to thermal environmental factors. Auth. doc. diss. Ryazan, 1950.
87. Fadeeva A.N. Creation of source material for pea breeding in the conditions of the northern part of the middle Volga region. // Author's Dissertation of Candidate of Biological Sciences.- SPb, 2001. C. 16.
88. Filimonova Yu.A. Biological features and productivity of bean varieties *(Phaeeolus vulgaris* L.) for vegetable use from the collection of VIR in the conditions of Krasnodar region. Author's thesis for the degree of Candidate of Agricultural Sciences. St.-Petersburg, 2009. p.6
89. Fedetov V.S. Pea. M., 1990, - P. 258.
90. Fursov D.I. Ability of bean seeds to repeated germination. - In book: Mater. Scientific Conference and Kharkov Agricultural Institute. Vyp.Z. Kharkov, 1968
91. Khalgildin V.V. Genetic bases of pea breeding. // Materials of the All-Union scientific and methodical meeting on pea breeding and genetics. - Ufa, 1971.- P.30-40.
92. Chekalin N.M., Korsakov N.I., Varlyakov M.D., Agarkova S.N. Golubev A.A., Goluvev A.I., Kurdin A.I., Lakhanov A.P. Selection of grain legume

crops. M.: Kolos, 1981.- P. 336.
93. Chesnokov V.A. New sources of raw materials for citric acid production. Bulletin of the State Leningrad University, No. 9, 1951.
94. Shlyakhturov D.S. Influence of method! S1vbya and norms of vis1vu nashnaya on the growth of KBacojii. In: Problems of Suchaenogo zemlya korystuvannaya: materiali nauk. conf. of young scientists. Ky1v: 2003. ECMO
95. www.ukradroconsult.com./.../30-novyh-zasuho ...
96. Asadova A.t. Botanizing and stydi of legumes. // International conference "Diversity, characterization and utilisation of plant genetic resources for enhanced resilience to climate chanje". Baku, Azerbaijan, 2011. P.155-156
97. Borojevic S., Willioms W.A. Genotype environment interaction in realisation to qreen area parameters, climatic faktors and qrain yield of Wheat. // Proc. 3 rd. Inter. Wheat Conf. Madrid, p. 38-52, 1980.
98. Beaver J.S., Kelly J.D. Yield compensation of bean genotypes growing in hillplots. Hort Sciensce. 1989. 24. 1. p. 137-138.
99. Finlayk K.W. and Wilkinson G.N.. The analysis of adaptatsion in plant breedinq programme. Austral. J. Agric. Res, 14, 1963. P.743-754.
100. Kapur, R; Satija, D.R.; Dahiya, B.C. Influence of plant population of agronomic traits in pigeon pea varieties. Agr. Sc. Dig. 1987. 7,3; p.159-162.
101. Oka J.I. Adaptabillty for seasons and locations and yield stabillty in crop varieties, its mechanism and selektsion. // Recent Adv.in Breeding (Tokyo),V.6.217 pp. 1975
102. Polignano G.B.,Uggent P., Perrino P.. Pattern analysis and genotypic x environment interactions in faba bean (Vicia faba L.) population. Euphytica 1989.40.1/2: p.31-41.
103. Matsuo T. Adaptabillty in Plants, With. Spesial reference to crop yeld. // Japan Comm. Inter. Biol. Proqram (Tokyo), V.6. 217 pp. 1975.
104. Wanjari K.B.. Variability and character Indian in blackqram (Vigna mungo). Indian J. agr. Sc. 1988, 58, 1; p. 48-51
105. file:///C://C:/Users/user-pc/Downloads/48-115-1SN%20(4).pdf

MIX
Papier aus verantwortungsvollen Quellen
Paper from responsible sources
FSC® C105338

Printed by Books on Demand GmbH, Norderstedt / Germany